Microarchitecture of VLSI Computers

NATO ASI Series

Advanced Science Institutes Series

A Series presenting the results of activities sponsored by the NATO Science Committee, which aims at the dissemination of advanced scientific and technological knowledge, with a view to strengthening links between scientific communities.

The Series is published by an international board of publishers in conjunction with the NATO Scientific Affairs Division

A	Life Sciences	Plenum Publishing Corporation
B	Physics	London and New York
C	Mathematical and Physical Sciences	D. Reidel Publishing Company Dordrecht and Boston
D	Behavioural and Social Sciences	Martinus Nijhoff Publishers Dordrecht/Boston/Lancaster
E	Applied Sciences	
F	Computer and Systems Sciences	Springer-Verlag Berlin/Heidelberg/New York
G	Ecological Sciences	

Series E: Applied Sciences – No. 96

Microarchitecture of VLSI Computers

edited by

P. Antognetti
University of Genova
Genova, Italy

F. Anceau
INPG
Grenoble, France

J. Vuillemin
INRIA
Paris, France

1985 **Martinus Nijhoff Publishers**
Dordrecht / Boston / Lancaster
Published in cooperation with NATO Scientific Affairs Division

Proceedings of the NATO Advanced Study Institute on Microarchitecture of VLSI Computers, Sogesta, Urbino, Italy, July 9–20, 1984

Library of Congress Cataloging in Publication Data

ISBN 90-247-3202-6 (this volume)
ISBN 90-247-2689-1 (series)

Distributors for the United States and Canada: Kluwer Boston, Inc., 190 Old Derby Street, Hingham, MA 02043, USA

Distributors for the UK and Ireland: Kluwer Academic Publishers, MTP Press Ltd, Falcon House, Queen Square, Lancaster LA1 1RN, UK

Distributors for all other countries: Kluwer Academic Publishers Group, Distribution Center, P.O. Box 322, 3300 AH Dordrecht, The Netherlands

PREFACE

We are about to enter a period of radical change in computer architecture. It is made necessary by advances in processing technology that will make it possible to build devices exceeding in performance and complexity anything conceived in the past. These advances, the logical extension of large - to very-large-scale integration (VLSI) are all but inevitable. With the large number of switching elements available in a single chip as promised by VLSI technology, the question that arises naturally is:
What can we do with this technology and how can we best utilize it? The final answer, whatever it may be, will be based on architectural concepts that probably will depart, in several cases, from past and present practices. Furthermore, as we continue to build increasingly powerful microprocessors permitted by VLSI process advances, the method of efficiently interconnecting them will become more and more important. In fact one serious drawback of VLSI technology is the limited number of pins on each chip. While VLSI chips provide an exponentially growing number of gates, the number of pins they provide remains almost constant. As a result communication becomes a very difficult design problem in the interconnection of VLSI chips. Due to the insufficient communication power and the high design cost of VLSI chips, computer systems employing VLSI technology will thus need to employ many architectural concepts that depart sharply from past and present practices.

This book focuses on the internal microarchitecture of present and future microprocessors and microcomputers in VLSI technology. The authors of the different chapters represent a well balanced blend of academic and industrial backgrounds. The academic authors describe more general, research-oriented topics, while the authors coming from industry describe in detail the architecture of their latest products. In this way the readers have a unique opportunity to obtain a full description of the present (industrial) and future (academic) architectural concepts for VLSI computers.

The content of this book results from the lecture notes handed out at the NATO ASI, held at SOGESTA, Urbino, Italy, from July 9 to

VI

July 20, 1984.

I would like to thank Drs. M. di Lullo and C. Sinclair of NATO's Scientific Affairs Division for their helpful assistance in organizing the ASI, and the lecturers for their timely and accurate presentations and for their careful preparation of the text.

P. Antognetti
Director of the NATO-ASI

University of Genova, Italy
January, 1985.

TABLE OF CONTENTS

VIII

PART I:
VLSI ARCHITECTURES FOR MICROPROCESSORS

MICRO/370

Richard W. Hadsell

IBM Thomas J. Watson Research Center
Yorktown Heights, New York 10598, U. S. A.

ABSTRACT.   Micro/370 will be a 32-bit single-chip NMOS micro-processor.   It directly implements 102 System/370 instructions and supports the execution of an additional 60 System/370 instructions by coprocessors. Separation of Control Space from 370 Space provides a means of emulating unimplemented instructions without consuming any of the user's memory space.   Its external bus is compatible with the Motorola MC68000 but supports unmultiplexed 32-bit addresses and 32-bit data.   The execution unit, bus controller, control store, instruction decoders, and clock phase generators are all on the chip. Microcode and functional specification of the execution unit are the direct result of Tredennick's Flowchart Method for hardware design. The Micro/370 execution unit contains two 32-bit data buses, a 32-bit adder, a 32-bit arithmetic and logic unit, a 64-bit shifter, and 2 sets of 16 32-bit general purpose registers.   The bus controller performs external bus arbitration, synchronizes external reset and interrupt signals, and runs bus cycles to read or write data in 1-byte, 2-byte, or 4-byte formats selected dynamically by the slave devices.   Microcode in a 64-Kbit control store controls the processor. Micro/370 runs nominally with a 20-MHz clock and a 100-ns processor cycle.

1   INTRODUCTION

1.1   The Micro/370 Project

Micro/370 is an IBM research project with two objectives:

- Document a method appropriate for designing a full-custom single-chip MOS microprocessor.

- Use the method to design a microprocessor implementation of the System/370 architecture.[1]

The Micro/370 microprocessor design is our example to illustrate the design method. I describe its architecture in this lecture. I will refer to the microprocessor design simply as "Micro/370". Documentation of the design method will be published elsewhere.

Three people are primarily responsible for Micro/370:

- Brion Shimamoto determined the subset of the System/370 architecture which Micro/370 would implement.

- Nick Tredennick designed the Processor — its microarchitecture, its microcode, and its logical design.

- Bruce D. Gavril specified the architecture of the Bus Controller — its microarchitecture and its appearance to the real (off-chip) world.

1.2 What Micro/370 Does

Micro/370 directly implements 102 System/370 instructions. Figure 1 compares the Micro/370 subset with the complete set of System/370 instructions. Of the 104 Standard Instructions, Micro/370 implements 91 of the General Instructions and Control Instructions but none of the I/O Instructions. Figure 2, Figure 3 , and Figure 4 list them. Micro/370 supports coprocessor execution of the 9 Decimal Instructions and the 51 Floating-Point and Extended-Precision Floating-Point Instructions. Figure 5 lists the remaining 11 instructions in the Micro/370 subset. Shimamoto estimates that Micro/370 will be able to execute its subset instructions at a rate of more than 250 kips (250,000 instructions per second) with a 10-MHz clock.

Micro/370 operates only in System/370 extended-control (EC) mode. Its System/370 program-status word (PSW) can not take on the basic-control (BC) mode format compatible with System/360. The Micro/370 PSW (see Figure 6) differs from the System/370 EC-mode PSW only in having a 32-bit instruction address, in bits 32-63.

Note: For the entire lecture, I will use IBM's bit notation: bits are numbered from left to right (most- to least-significant) starting with 0. I will point out any exception.

---

[1]   I assume for this lecture a general knowledge of IBM's System/370 architecture. It is described thoroughly in Reference (1).

5

Comparing System/370 to Micro/370:
Total Number of Instructions by Facility

<u>System/370:Micro/370</u>

- Universal Instruction Set 157:91
  - Commercial Instruction Set. 113:91
    - Standard Instruction Set. 104:91
      - General Instructions. 87:85
      - Control Instructions. 9:6
      - I/O Instructions. 8:0
    - Decimal Instruction Set 9:0
  - Floating Point Facility 44:0
- Advanced Control Program Support Feature. 4:4
  - Conditional Swapping Facility 2:2
  - PSW-Key Handling Facility 2:2
- Branch And Save Facility. 2:2
- Channel-Set Switching Facility. 2:0
- CPU Timer and (TOD) Clock Comparator Facility 4:0
- Direct Control Facility 2:0
- Dual-Address Space (DAS) Facility 12:0
- ECPS:VM/370 - Extended Control-Program Support for VM/370 ** 22:0
- ECPS:VS1 - Extended Control-Program Support for OS/VS1. . ** 39:0
- Extended Facility and 3033 Extension (3033X) Feature. . . .15:0
  - IPTE and TPROT instructions 2:0
  - ECPS:MVS - Extended Control-Program Support for MVS **12:0
  - ADD FRR instruction . . . . ** 1:0
- Extended-Precision Floating Point Facility. 7:0
- Move Inverse Facility 1:1
- Multiply-Add Facility 1:0
- Multiprocessing (MP) Facility 4:0
- Recovery Extensions Facility. 1:0
- Storage-Key-Instruction Extensions Facility 3:0
- Suspend and Resume Facility 1:0
- Test Block Facility 1:0
- Translation Facility. 5:4

Total: 283:102
** Control-Program Dependent: - 74:0
Control-Program Independent: 209:102

Figure 1. Comparison of Micro/370 with System/370 instructions.

Whenever Micro/370 loads any of PSW bits 1, 5-11, 13-16, and 20-23, it reports the new PSW to the real (off-chip) world with a special bus cycle. Although Micro/370 ignores the state of some of those bits, attached devices (like a virtual address translator) can be controlled by them.

The Micro/370 Bus Controller interfaces the Micro/370 Processor with the real world. It supports bus access protocols which are compatible with the Motorola M68000 component bus. That is, Micro/370 can communicate with 16-bit MC68000 devices and 8-bit

6

Instructions Implemented in Micro/370

• General Instructions		Op Code
Add	(AR):	1A
Add	(A):	5A
Add Halfword	(AH):	4A
Add Logical	(ALR):	1E
Add Logical	(AL):	5E
AND	(NR):	14
AND	(N):	54
AND	(NI):	94
AND	(NC):	D4
Branch and Link	(BALR):	05
Branch and Link	(BAL):	45
Branch on Condition	(BCR):	07
Branch on Condition	(BC):	47
Branch on Count	(BCTR):	06
Branch on Count	(BCT):	46
Branch on Index High	(BXH):	86
Branch on Index Low or Equal	(BXLE):	87
Compare	(CR):	19
Compare	(C):	59
Compare Halfword	(CH):	49
Compare Logical	(CLR):	15
Compare Logical	(CL):	55
Compare Logical	(CLI):	95
Compare Logical	(CLC):	D5
Compare Logical Characters under Mask	(CLM):	BD
Compare Logical Long	(CLCL):	0F
Convert to Binary	(CVB):	4F
Convert to Decimal	(CVD):	4E
Divide	(DR):	1D
Divide	(D):	5D
Exclusive OR	(XR):	17
Exclusive OR	(X):	57
Exclusive OR	(XI):	97
Exclusive OR	(XC):	D7
Execute	(EX):	44
Insert Character	(IC):	43
Insert Characters under Mask	(ICM):	BF
Load	(LR):	18
Load	(L):	58
Load Address	(LA):	41
Load and Test	(LTR):	12
Load Complement	(LCR):	13

Figure 2. System/370 Standard Instruction Set (part 1 of 3).

Instructions Implemented in Micro/370 (cont.)

• General Instructions (cont.)   Op Code

Load Halfword	(LH):	48
Load Multiple	(LM):	98
Load Negative	(LNR):	11
Load Positive	(LPR):	10
Move	(MVI):	92
Move	(MVC):	D2
Move Long	(MVCL):	0E
Move Numerics	(MVN):	D1
Move with Offset	(MVO):	F1
Move Zones	(MVZ):	D3
Multiply	(MR):	1C
Multiply	(M):	5C
Multiply Halfword	(MH):	4C
OR	(OR):	16
OR	(O):	56
OR	(OI):	96
OR	(OC):	D6
Pack	(PACK):	F2
Set Program Mask	(SPM):	04
Shift Left Double	(SLDA):	8F
Shift Left Double Logical	(SLDL):	8D
Shift Left Single	(SLA):	8B
Shift Left Single Logical	(SLL):	89
Shift Right Double	(SRDA):	8E
Shift Right Double Logical	(SRDL):	8C
Shift Right Single	(SRA):	8A
Shift Right Single Logical	(SRL):	88
Store	(ST):	50
Store Character	(STC):	42
Store Characters under Mask	(STCM):	BE
Store Halfword	(STH):	40
Store Multiple	(STM):	90
Subtract	(SR):	1B
Subtract	(S):	5B
Subtract Halfword	(SH):	4B
Subtract Logical	(SLR):	1F
Subtract Logical	(SL):	5F
Supervisor Call	(SVC):	0A
Test under Mask	(TM):	91
Test and Set	(TS):	93
Translate	(TR):	DC
Translate and Test	(TRT):	DD
Unpack	(UNPK):	F3

Figure 3.   System/370 Standard Instruction Set (part 2 of 3).

Instructions Implemented in Micro/370 (cont.)

- Control Instructions   Op Code

    Diagnose   :   83
    Insert Storage Key   (ISK):   09
    Load PSW   (LPSW):   82
    Set Storage Key   (SSK):   08
    Set System Mask   (SSM):   80
    Store CPU ID   (STIDP):   B202

Instructions Not Implemented in Micro/370

- General Instructions   Op Code

    Monitor Call   (MC):   AF
    Store Clock   (STCK):   B205

- Control Instructions   Op Code

    Load Control   (LCTL):   B7
    Set Clock   (SCK):   B204
    Store Control   (STCTL):   B6

- I/O Instructions   Op Code

    Clear I/O   (CLRIO):   9D01
    Halt Device   (HDV):   9E01
    Halt I/O   (HIO):   9E00
    Start I/O   (SIO):   9C00
    Start I/O Fast Release   (SIOF):   9C01
    Test Channel   (TCH):   9F00
    Test I/O   (TIO):   9D00
    Store Channel ID   (STIDC):   B203

Figure 4. System/370 Standard Instruction Set (part 3 of 3).

MC68008 devices.[2] However, the Micro/370 data bus is 32 bits wide, and the address bus is also 32 bits wide. The Micro/370 supports any mixture of attached 32-bit, 16-bit, and 8-bit devices, including M6800 peripherals. It determines dynamically, based on the external device's response, how much data have been transferred. If necessary, it runs additional bus cycles to transfer the remaining data.

Just as with the M68000 component bus, a Function Code separates Data address space from Program space and User space from Supervisor space, providing four distinct memory spaces, but we call User space "370 Space" and Supervisor space "Control Space." Micro/370 commu-

---

[2]   I use "device" to refer to any memory or peripheral which at-
      taches to the Micro/370 Bus.

Non-Universal System/370 Facilities

- Instructions Implemented in Micro/370

  - Advanced Control Program Support Feature

    - Conditional Swapping Facility   Op Code

      Compare and Swap   (CS):   BA
      Compare Double and Swap   (CDS):   BB

    - PSW-Key Handling Facility   Op Code

      Insert PSW Key   (IPK):   B20B
      Set PSW Key from Address   (SPKA):   B20A

    - Branch and Save Facility   Op Code

      Branch and Save   (BASR):   0D
      Branch and Save   (BAS):   4D

    - Move Inverse Facility   Op Code

      Move Inverse   (MVCIN):   E8

    - Translation Facility   Op Code

      Purge TLB   (PTLB):   B20D
      Reset Reference Bit   (RRB):   B213
      Store Then AND System Mask (STNSM):   AC
      Store Then OR System Mask (STOSM):   AD

- Instructions Not Implemented in Micro/370

  - Translation Facility   Op Code

    Load Real Address   (LRA):   B1

  - All other facilities not listed above

Figure 5. System/370 Instructions not in the Universal Instruction Set.

nicates with special devices with special bus cycles called "Service Cycles," identified by a special Function Code. Micro/370 handles standard System/370 interruptions as well as the 7 M68000 interrupts.

Micro/370 does not execute the 51 floating-point instructions or the 9 decimal instructions of System/370, but it can let them be executed by attached coprocessors (one for floating-point, a different one for decimal instructions). Micro/370 does all instruc-

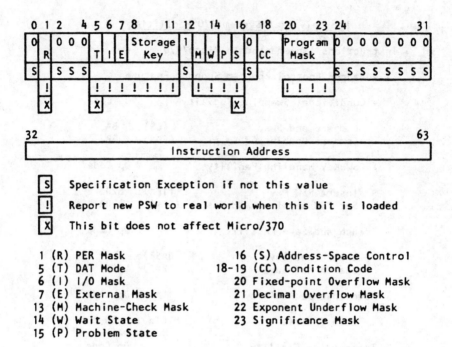

Figure 6. Micro/370 PSW.

tion fetching, decodes the instruction, and determines whether the coprocessor is attached. If it is, Micro/370 does whatever the coprocessor needs for that instruction (possibly fetching operands) and triggers the coprocessor with a Service Cycle. Micro/370 then uses a Service Cycle to wait for the coprocessor's completion and finishes the instruction execution by setting the System/370 Condition Code or by processing an exception.

To make it easier for a system to support programs containing unimplemented instructions, Micro/370 has a Dual Mode Feature. This consists of logic to support a private memory, 16 additional (32-bit) general registers, and a set of DIAGNOSE instructions. The private memory can be used as a "writeable control store." The instructions in this "Control Space" are System/370 instructions. The DIAGNOSE instructions support the concept of these instructions being micro-code, relative to instructions executing from the normal "370 Space." With several restrictions, instructions fetched from the Control Space may operate on data in either Control or 370 Space.

1.3 The Micro/370 Chip Floorplan

The Micro/370 chip is 10 mm x 10 mm. Shimamoto estimates that it has 140000 device sites. Figure 7 shows the chip floorplan, in-

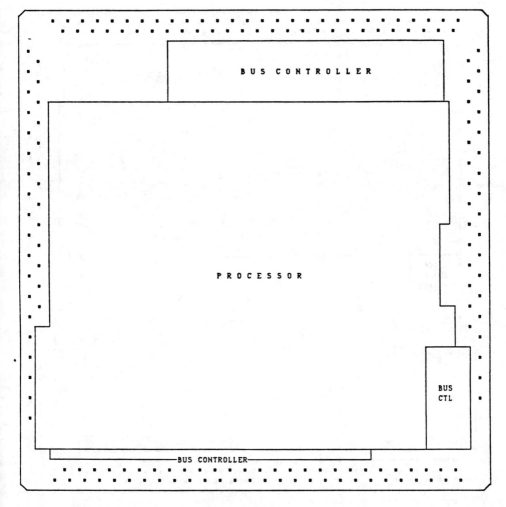

10mm x 10mm

Figure 7. Micro/370 Floorplan.

dicating the areas dedicated to the Processor and the Bus Controller.
Figure 8 shows more floorplan detail. The Processor is divided into
Clock-Phase Generator, State Machine, and Execution Unit. The Ad-
dress Bus Control and Data Bus Control are below and to the right
of the Execution Unit, distinct from the main portion of the Bus
Controller.

Figure 9 shows enough detail in the State Machine and the Execution Unit (E-Unit) to understand the Micro/370 microarchitecture
at a high level. State Machine Control determines a 10-bit Control
Store Address (CSA) for the next state. The control store consists

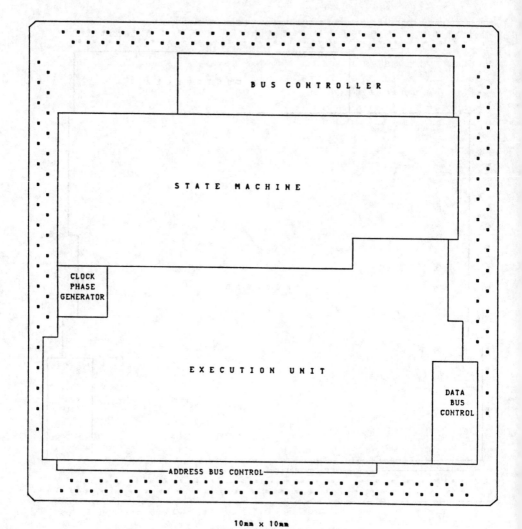

10mm x 10mm

Figure 8.   Micro/370 Floorplan.

of two ROMs with a total storage capacity of about 62K bits.   The
Microrom uses the high-order 6 bits of the new CSA to select a row
of 16 18-bit Microwords.   Simultaneously, the Nanorom uses the
high-order 8 bits of the new CSA to select a row of 4 72-bit
Nanowords.   The low-order 4 CSA bits select the Microword for the
next state from the Microrom output multiplexer.   Similarly, the
low-order 2 CSA bits select the Nanoword for the next state from the
Nanorom output multiplexer.

The Microword is decoded by PLAs which drive both the Bus Controller, via the Processor Command Register (PCR), and State Machine

Control (next-address selection). The Nanoword is decoded by the
E-Unit Control PLAs (which drive the E-Unit). This separation between Microword and Nanoword in what might otherwise be a single
microcode word makes possible a major saving in control store space.
Several states of the Processor, each having a distinct Microword,
can require exactly the same functions of the E-Unit; that is, they
can share a single Nanoword. The sharing takes place by assigning
the several states to addresses which differ only with respect to
specific bits. The Nanorom ignores those bits for this group of
addresses and always selects the same row of 4 Nanowords, no matter
which of the several states is selected by the CSA. This scheme

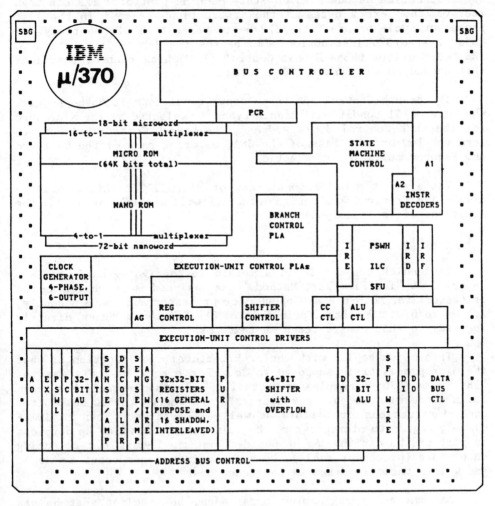

10mm x 10mm

Figure 9. Micro/370 Floorplan.

reduces the Nanorom space requirement from 70K bits to about 44K bits.

The E-Unit sends external bus addresses via its AO (Address Out) register to Address Bus Control and external data via its DO (Data Out) Register to Data Bus Control. It accepts external data via its DI (Data In) Register from Data Bus Control.

A Special Function Unit, which sits just above the E-Unit, contains among other things 3 16-bit instruction prefetch registers. IRF receives 2 bytes of the instruction stream from Data Bus Control. Under Nanoword control, IRF passes its 2 bytes to IRD, which drives the Instruction Decode PLAs in State Machine Control. IRF can then receive the next 2 bytes of the instruction stream. In the last processor cycle before Micro/370 begins execution of the newly decoded System/370 instruction, IRD passes the 2-byte instruction to IRE. IRE drives those E-Unit Control PLAs which control functions that depend on the System/370 instruction.

The Branch Control PLA takes input conditions from the E-Unit. When one of 31 conditional branch types is selected by the Microword for the next Control Store Address, the Branch Control PLA outputs form the low-order 2 bits of the CSA, becoming part of the Microrom and Nanorom multiplexer selection.

These are the major components of Micro/370. After briefly discussing Tredennick's design method, I will describe in detail the functions of these components.

1.4 The Flowchart Method

Tredennick has developed a method for designing hardware (2). He calls it "The Flowchart Method." He has used it in designing the processors for Motorola's MC68000 microprocessor and, more recently, for Micro/370. The basic principle of The Flowchart Method directly contradicts what others teach in textbooks.

Tredennick begins with the architectural specification of what the microprocessor is supposed to do. For Micro/370, this was the "IBM System/370 Principles of Operation." He then establishes goals which guide his design. For Micro/370, he wanted the 10 most often used instructions to perform as well as possible. Then he immediately begins to microprogram. He uses a notation which he develops as part of the design. As he decides what the E-Unit must do during each state (processor cycle), he hypothesizes a component to do it and writes terms to describe the functions.

As his microprogramming progresses, he examines what he has done and summarizes the functions of the components. He may decide that the creation of new components or new functions would improve the processor's performance. He may decide that one of his creations is not useful after all, or he may learn from the circuit designers that the component or function is too hard to build in the chosen technology. He then throws out the component or function and achieves the result some other way.

When the lists of functions seem complete, he establishes a format for the microcode and draws a block diagram of the processor. This can be a year or more after the start of the design. That is, the microarchitecture of the processor is an output of the design process, which included the microprogramming. This is the opposite of design methods which begin the design process by picking (in some undefined way, perhaps based on popular ideas like "On-chip caches improve performance") the processor's configuration, controller, and microcode format. Tredennick never deals with those issues until a major portion of the microprogramming is done. By then, he can base the decisions on what is really needed to achieve his goals. He does not need to guess the effects of design alternatives.

To illustrate Tredennick's Flowchart style, Figure 10 shows a small section of the Micro/370 "Flowcharts." This section includes 3 sequences of states. The sequence on the left processes the BCR instruction when the register specified in the R2 field is not register 0. The sequence in the middle column processes the BCR instruction when R2 is register 0 — in effect, a NO-OP instruction. The sequence on the right processes the BCTR instruction when the R2 register is not register 0. The information at the start of each sequence has no effect on the microcode; it is included as commentary to make it easier to read the flowcharts and as an aid in designing the chip's decoders which select the sequence.

Each large box in a sequence corresponds to a state and is associated with a Microword and a Nanoword. The small boxes within each state identify functions which the Processor must perform or provide the designer with useful comments. Item 1 is the name of the state. Items 2 are functions of the E-Unit. Item 3 is the "ALU Column," which selects an arithmetic or logic function for the ALU. Item 4 specifies a direct (unconditional) branch to the state it names. Item 5 identifies a conditional branch, and items 6 show the possible conditions for this particular form of conditional branch and the names of the target states. Item 7 indicates that the next state must be selected by the A1 Instruction Decode PLA, according to the System/370 instruction in IRD. Item 8 is the name of the state whose Nanoword is used by this state (the two states share it); the E-Unit functions are shown here in the hardcopy only to help the designer. Item 9 specifies a Bus Access request, which will be made to the Bus Controller via the PCR. Item 10 gives the data width for the Bus Access. Item 11 shows, as another aid to the designer, how many other states share this state's Nanoword. Item 12 tells the

Micro/370 Flowcharts: FC SCRIPT M370   last updated 84/04/23 18:58:13

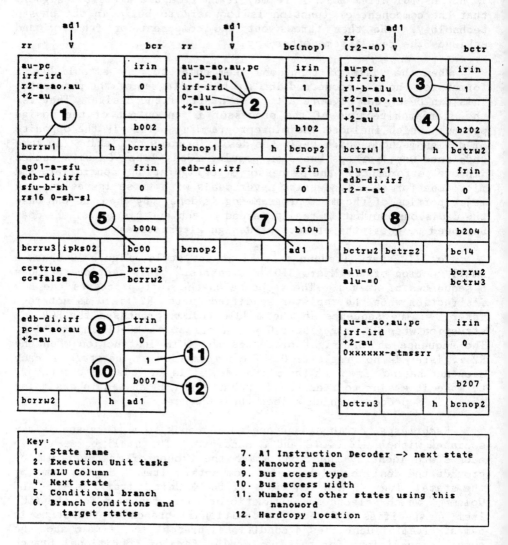

Key:
1. State name
2. Execution Unit tasks
3. ALU Column
4. Next state
5. Conditional branch
6. Branch conditions and
   target states
7. A1 Instruction Decoder —> next state
8. Nanoword name
9. Bus access type
10. Bus access width
11. Number of other states using this
    nanoword
12. Hardcopy location

Figure 10. A section of Micro/370 Flowcharts.

program which prints the hardcopy where to place this state (page b, column 0, row 7).

2 EXECUTION UNIT

In this section I describe the functions of each component of
the Micro/370 Execution Unit (E-Unit). The chip floorplan in
Figure 9 shows roughly their placement and the area each occupies.
Figure 11 shows the data paths which interconnect them.

2.1 Buses

Two 32-bit data buses (named 'A' and 'B') span the E-Unit. Each
has two sections ('P' and 'D') which can be coupled when needed for
a particular data transfer. These buses provide paths for up to four
data transfers concurrently in a single processor cycle. To eliminate the need for large drivers in each component, the buses have
sense amplifiers. They sense when a source is attempting to drive
a bus line and amplify the signal to drive as many as 3 components.
The P and D sections have separate sense amplifiers, for use when
they are not coupled. When they are coupled, only the D sense amplifiers turn on and drive the entire bus.

The sense amplifiers on the D bus sections provide a second
function. The flowcharts can select which of the amplifiers will
turn on during the data transfer. The selection is arranged in
bytes. When the amplifiers for one or more of the 4 bytes of data
do not turn on, no transfer takes place. The E-Unit components are
designed so that, when the amplifiers do not drive the bus lines for
a particular byte, their registers retain their old contents. This

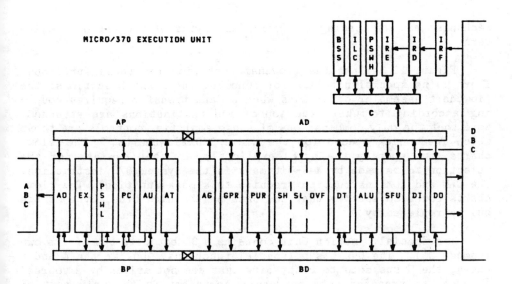

Figure 11. Micro/370 Execution Unit data paths.

When the Flowcharts say:	Then these sense amplifiers turn on:				
	ad(0:7)	ad(8:15)	ad(16:23)	ad(24:31)	ap(0:31)
-a-	X	X	X	X	
-am-	M0	M1	M2	M3	
-az-	Z	X	X	X	
-a2-			X		
-a7-		X	X	X	
Do not couple ap-ad					X
	bd(0:7)	bd(8:15)	bd(16:23)	bd(24:31)	bp(0:31)
-b-	X	X	X	X	
-bm-	M0	M1	M2	M3	
-bz-	Z	X	X	X	
-b1-				X	
-b2-			X		
-b7-		X	X	X	
-be-	X	X	X		
Do not couple bp-bd					X

Notes:
1. X: turn on sense amps
2. M0: turn on sense amps only if ire(12) = 1
   M1: turn on sense amps only if ire(13) = 1
   M2: turn on sense amps only if ire(14) = 1
   M3: turn on sense amps only if ire(15) = 1
3. Z: drive bus lines with 0

Figure 12. E-Unit Bus Sense Amplifier functions.

mechanism allows loading of individual bytes of the various registers.

Figure 12 defines the flowchart notation for these functions.
There is no special notation for coupling the P and D sections: the
flowchart assembler determines when a data transfer requires coupling according to where the source and destinations are attached.
Note that the only combinations of sense amplifier controls are those
which Tredennick needed for the Micro/370 Flowcharts.   The flowcharts can specify the bytes explicitly, or they can ask for the
selection to be made by a 4-bit mask in the System/370 instruction.
The CLM and ICM instructions require this mask function. The flowcharts can also specify that the data on the high-order byte of the
bus be replaced by 0.

The Special Function Unit contains a 32-bit data bus of its own
(named 'C'). The bus has no special functions. Unlike the A and B
buses, the C bus sets to 0 any bits that are not driven by a source.
In order to transfer data between a component in the main part of
the E-Unit and a component in the Special Function Unit, the data

When the Flowcharts say:	Then:
a-ao	ap(0:31) → ao(0:31)
au-ao	auo(0:31) → ao(0:31)
b-ao	bp(0:31) → ao(0:31)

Figure 13. AO functions.

must first go to the 32-bit SFU Register and, in a later processor cycle, to its ultimate destination.

2.2 AO

The 32-bit AO Register is at the left end of the E-Unit. It supplies addresses to Address Bus Control (ABC) for external bus accesses. For each bus access requested by the Processor, it receives a new value from the A bus, the B bus, or AU (see Figure 13).

2.3 EX

Micro/370 uses the 32-bit EX Register for processing the EX instruction. It loads the contents of AU (which contains the EX instruction address plus 4) into EX before executing the target instruction, and it turns on the EB (Execute Bit) control flag. It then fetches the target instruction, modifies it as necessary, and sets up PC and AU to look as though the target had been part of the normal instruction stream. When Micro/370 finishes processing the target instruction, the EB flag causes processing to return to the EX instruction, and the contents of the EX Register are used as the next address in the instruction stream.

When the Flowcharts say:	And:	Then:
b-ex		bp(0:31) → ex(0:31)
ex-a		ex(0:31) → ap(0:31)
ex-b		ex(0:31) → bp(0:31)
pce-a	eb = 1	ex(0:31) → ap(0:31)
pce-b	eb = 1	ex(0:31) → bp(0:31)
Notes: 1. eb: Execution Bit is on during execution of the target instruction of an EX instruction.		

Figure 14. EX functions.

Certain instructions, including BAL, BAS, and SVC, behave differently when they are the target of an EX instruction. Part of their execution results involve an address which normally points to the next instruction. When they are the target of an EX instruction, however, the address must point to the location following the EX instruction, NOT the location following the target instruction. The flowcharts then refer to 'PCE', which is interpreted as 'PC' if EB is off and 'EX' if EB is on. Figure 14 shows the functions of the EX Register.

2.4 PSWL

PSWL is related to the low-order half (32 bits) of the System/370 Program Status Word. Unlike the standard System/370 PSW, which has only a 24-bit Instruction Address, the entire 32 bits of PSWL are an address. When Micro/370 begins processing a System/370 instruction, PSWL receives from AU the current instruction's location plus 4. This is the only way that PSWL is ever loaded (see Figure 15). Whenever the real world needs to see the PSW (e.g., as the result of a PSW swap), Micro/370 adjusts the address according to the known offset of the contents of PSWL and the length of the current instruction.

2.5 PC

Micro/370 uses the 32-bit PC Register as a Program Counter. When Micro/370 starts processing a System/370 instruction, the contents of PC point to the instruction's address plus 2. As Micro/370 prefetches the following instruction, it updates PC. When the flowcharts say 'PCE' and EB (the Execution Bit) is off, they refer to PC. (See section "2.3 EX".) Figure 16 lists the functions of PC.

2.6 AU

AU is a 32-bit adder. (It can also subtract.) Figure 17 lists its functions. The AU result register, named 'AUO', is latched whenever AU calculates a new value. The carry out of the most-

When the Flowcharts say:	Then:
a1	auo(0:31) → pswl(0:31)
pswl-b	pswl(0:31) → bp(0:31)
Notes: 1. 'a1': start executing a new S/370 instruction.	

Figure 15. PSWL functions.

When the Flowcharts say:	And:	Then:
a-pc		ap(0:31) → pc(0:31)
au-pc		auo(0:31) → pc(0:31)
b-pc		bp(0:31) → pc(0:31)
pc-a		pc(0:31) → ap(0:31)
pc-b		pc(0:31) → bp(0:31)
pce-a	eb = 0	pc(0:31) → ap(0:31)
pce-b	eb = 0	pc(0:31) → bp(0:31)
Notes: 1. eb: Execution Bit is on during execution of the target instruction of an EX instruction.		

Figure 16. PC functions.

significant bit is also latched. AU drives four condition signals:
AUC, AUN, and AUZ are the carry, negative, and zero flags for the
full 32 bits; AUZS is the zero flag for the low-order byte of the
result.   The Branch Control PLA uses these four conditions in re-
solving certain microprogram branch types.

Micro/370 uses AU for calculating addresses and whenever ALU
is already busy with another task. When Micro/370 starts processing
a System/370 instruction, the contents of AUO point to the instruction's address plus 4.   This is the location of the next
halfword to be fetched in the instruction stream, if no branch is
taken.

2.7 AT

AT is a 32-bit register.   Figure 18 lists its functions.
Micro/370 uses it frequently as a temporary register, especially for
addresses.

2.8 AG

AG generates 32-bit constants.   It can provide source data for
either the A bus or the B bus, but not both in a single processor
cycle.   Figure 19 lists the possible constants. Many of them are
addresses fixed in System/370 architecture for swapping PSWs or
storing interruption codes.

2.9 GPR

System/370 specifies 16 32-bit General Purpose Registers.
Micro/370 has an additional set of 16 32-bit registers, called Shadow
Registers.   Shadow Registers are part of the Dual Mode Feature.

When the Flowcharts say:	Then:
au-a	auo(0:31) → ap(0:31)
au-b	auo(0:31) → bp(0:31)
a-au	ap(0:31) → aua(0:31)
b-au	bp(0:31) → aub(0:31)
0-au	0 → aub(0:31)
+1-au	0 → aub(0:28); 001 → aub(29:31)
+2-au	0 → aub(0:28); 010 → aub(29:31)
+3-au	0 → aub(0:28); 011 → aub(29:31)
+4-au	0 → aub(0:28); 100 → aub(29:31)
+6-au	0 → aub(0:28); 110 → aub(29:31)
+7-au	0 → aub(0:28); 111 → aub(29:31)
-1-au	1 → aub(0:28); 111 → aub(29:31)
-2-au	1 → aub(0:28); 110 → aub(29:31)
-3-au	1 → aub(0:28); 101 → aub(29:31)
-4-au	1 → aub(0:28); 100 → aub(29:31)
-7-au	1 → aub(0:28); 001 → aub(29:31)
-8-au	1 → aub(0:28); 000 → aub(29:31)
Notes: 1. auo(0:31): AU result register 2. aua(0:31): AU input A aub(0:31): AU input B 3. AU usually adds: A + B → auo(0:31). However, if the Flowcharts say 's', AU subtracts: B - A → auo(0:31).	

Figure 17. AU functions.

GPR functions are controlled in two stages. As Figure 20 shows, the first stage maps Nanoword fields ("What the Flowcharts say") and the System/370 instruction fields into a 4-bit register number and a selection of the data path. System/370 instruction format defi-

When the Flowcharts say:	Then:
a-at	ap(0:31) → at(0:31)
b-at	bp(0:31) → at(0:31)
at-a	at(0:31) → ap(0:31)
at-b	at(0:31) → bp(0:31)

Figure 18. AT functions.

0	10	BA	80000
1	20	E8	440000
2	28	F0	3700400
3	30	F2	66666666
6	40	FA	B1000000
7	80	FE	B2200000
8	86	FF	B80840FF
9	88	100	FF000000
C	8C	3F00	FFF00000
F	B8	FFOF	FFFF8000

Notes:
1. Values are shown in hex.
2. Output data are right-justified with leading 0s.

Figure 19. AG constants.

nitions call the instruction fields which designate a register 'R1',
'R2', 'R3', 'X2', 'B1', and 'B2'. The Micro/370 Flowcharts call
those same fields 'R1', 'R2', 'R2', 'X2', 'RB', and 'RB', respectively. The RB register field is part of the second or third
halfword of the instruction; it is the base register for an operand

When the Flowcharts say:	Then Register Control says:	And:
a-r1	ire(8:11) → ra(0:3)	a → r
a-r2	ire(12:15) → ra(0:3)	a → r
a-r1(1)	ire(8:10) → ra(0:2); 1 → ra(3)	a → r
a-r2(1)	ire(12:14) → ra(0:2); 1 → ra(3)	a → r
a-re1	0001 → ra(0:3)	a → r
r1-a	ire(8:11) → ra(0:3)	r → a
r2-a	ire(12:15) → ra(0:3)	r → a
r1(1)-a	ire(8:10) → ra(0:2); 1 → ra(3)	r → a
r2(1)-a	ire(12:14) → ra(0:2); 1 → ra(3)	r → a
rb-a	di(16:19) → ra(0:3)	(r|0) → a
x2-a	ire(12:15) → ra(0:3)	(r|0) → a
b-r1	ire(8:11) → rb(0:3)	b → r
b-r2	ire(12:15) → rb(0:3)	b → r
b-r1(1)	ire(8:10) → rb(0:2); 1 → rb(3)	b → r
b-r2(1)	ire(12:14) → rb(0:2); 1 → rb(3)	b → r
b-re2	0010 → rb(0:3)	b → r
r1-b	ire(8:11) → rb(0:3)	r → b
r2-b	ire(12:15) → rb(0:3)	r → b
r1(1)-b	ire(8:10) → rb(0:2); 1 → rb(3)	r → b
r2(1)-b	ire(12:14) → rb(0:2); 1 → rb(3)	r → b
rb-b	di(16:19) → rb(0:3)	(r|0) → b
x2-b	ire(12:15) → rb(0:3)	(r|0) → b

Figure 20. GPR Control — Stage 1.

address. When Micro/370 refers to RB, the appropriate halfword is
in the low-order half of the DI register, so DI(16:19) are the 4 bits
specifying the base register. All other register fields are in IRE,
which contains the first halfword of the instruction.

X2 and RB indicate index and base registers for an address.
However, register 0 is never used as an index or base register. If
X2 or RB is 0, the value '0' is the index or base address.
Figure 20 uses the notation '(r|0)' to specify this alternative.
The flowchart notation 'R1(1)' or 'R2(1)' selects the odd register
of an even-odd register pair.

The second stage of GPR control (Figure 21) translates the
register and data path selection of the first stage into one or two
data transfer selections, according to the state of two control
flags. The 'RRS' flag selects either a General Purpose Register or
a Shadow Register as the source for a data transfer from GPR to a
bus. The 'WRS' flag selects either a General Purpose Register or

When Register Control says:	And:	And:	Then:
a → r		wrs = 0	ad → r<ra>
a → r		wrs = 1	ad → r<ra>; ad → s<ra>
r → a		rrs = 0	s<ra> → ad
r → a		rrs = 1	r<ra> → ad
(r|0) → a	ra = 0		0 → ad
(r|0) → a	ra ≠ 0	rrs = 0	s<ra> → ad
(r|0) → a	ra ≠ 0	rrs = 1	r<ra> → ad
b → r		wrs = 0	bd → r<rb>
b → r		wrs = 1	bd → r<rb>; bd → s<rb>
r → b		rrs = 0	s<rb> → bd
r → b		rrs = 1	r<rb> → bd
(r|0) → b	rb = 0		0 → bd
(r|0) → b	rb ≠ 0	rrs = 0	s<rb> → bd
(r|0) → b	rb ≠ 0	rrs = 1	r<rb> → bd

Notes:
1. rrs: Read Registers control flag
   wrs: Write Registers control flag
2. r<ra>: general purpose register number ra(0:3)
   s<ra>: shadow register number ra(0:3)
   r<rb>: general purpose register number rb(0:3)
   s<rb>: shadow register number rb(0:3)

Figure 21. GPR Control — Stage 2.

both a General Purpose Register and a Shadow Register as the destination(s) for a data transfer to GPR from a bus.

2.10 PUR

Micro/370 uses the 8-bit Pack-Unpack (PUR) Register for instructions which manipulate 4-bit nibbles of data. Figure 22 lists its functions. Although PUR only stores a byte of data, it generates a complete 32-bit word when it is the source for a bus transfer.

2.11 Shifter

The Micro/370 Shifter can shift 64 bits of data 0 to 63 bits in either direction in one processor cycle. It latches its output in two 32-bit registers named 'SH' (the high-order word) and 'SL' (the low-order word). Input data to the high half of the Shifter come either from the B bus or from SH; input to the low half come either from the A bus or from SL. Figure 23 lists the complete range of Shifter functions. The Shifter can operate on the entire 64 bits or on the high-order word alone. The functions include a variety of carry-in bits and data-wrap connections.

2.12 DT

DT is a 32-bit register. Figure 24 lists its functions. Micro/370 uses it frequently as a temporary register.

When the Flowcharts say:	For instruction(s):	Then:
a-pur	MVO,PACK	ad(28:31) → pur(0:3)
	CVB,UNPK	ad(24:31) → pur(0:7)
	others	ad(24:27) → pur(0:3)
b-pur	PACK	0 → pur(0:3); bd(28:31) → pur(4:7)
	others	bd(28:31) → pur(4:7)
pur-a	UNPK	0 → ad(0:23); 1 → ad(24:27); pur(4:7) → ad(28:31)
	others	0 → ad(0:23); pur(0:7) → ad(24:31)
purh-a	all	0 → ad(0:27); pur(0:3) → ad(28:31)
purl-a	all	0 → ad(0:27); pur(4:7) → ad(28:31)
pur-b	UNPK	0 → bd(0:23); 1 → bd(24:27); pur(0:3) → bd(28:31)
	others	0 → bd(0:23); pur(0:7) → bd(24:31)

Figure 22. PUR functions.

When the	Then the Shifter does this:		
Flowcharts say:	Shift	Carry in	Connect
ls1 sh-0	sh left 1	sh(31) ← 0	
ls1 sh-sl-0	sh,sl left 1	sl(31) ← 0	sh(31) ← sl(0)
ls1 sh-sl-1	sh,sl left 1	sl(31) ← 1	sh(31) ← sl(0)
ls1 sl-sh-0	sh,sl left 1	sh(31) ← 0	sl(31) ← sh(0)
ls2 sl-sh-0	sh,sl left 2	sh(30:31) ← 0	sl(30:31) ← sh(0:1)
ls4 sh-sl-0	sh,sl left 4	sl(28:31) ← 0	sh(28:31) ← sl(0:3)
ls8 sh-0	sh left 8	sh(24:31) ← 0	
ls8 sh-sl-0	sh,sl left 8	sl(24:31) ← 0	sh(24:31) ← sl(0:7)
ls16 sh-0	sh left 16	sh(16:31) ← 0	
rs1 0-sh-sl	sh,sl right 1	0 → sh(0)	sh(31) → sl(0)
rs1 n.eor.v- sh-sl	sh,sl right 1	(alun XOR aluv) → sh(0)	sh(31) → sl(0)
rs8 0-sh	sh right 8	0 → sh(0:7)	
rs8 0-sh-sl	sh,sl right 8	0 → sh(0:7)	sh(24:31) → sl(0:7)
rs16 0-sh-sl	sh,sl right 16	0 → sh(0:15)	sh(16:31) → sl(0:15)
rs32 0-sh-sl		0 → sh(0:31)	sh(0:31) → sl(0:31)
shift	ALU contains shift amount (0-63). IRE selects Left/Right, Single/Double, Arithmetic/Logical.		

Notes:
1. sh: high-order 32 bits of Shifter
   sl: low-order 32 bits of Shifter
2. Input for sh(0:31) comes from bd(0:31) or previous sh(0:31).
   Input for sl(0:31) comes from ad(0:31) or previous sl(0:31).
3. alun and aluv: ALU word condition flags (negative and overflow)

Figure 23. Shifter functions.

2.13 ALU

The Micro/370 Arithmetic-Logic Unit (ALU) provides a variety of arithmetic and logic functions (see Figure 25), selected by the flowchart ALU Column choice and by the System/370 instruction in IRE. ALU Columns 1 and 4 are Add and Subtract functions for all instructions; they fulfill a wide variety of purposes. Columns 5 and 6 are And and Exclusive-Or functions; they support new-PSW checking

When the Flowcharts say:	Then:
a-dt	ad(0:31) → dt(0:31)
alu-dt	aluo(0:31) → dt(0:31)
b-dt	bd(0:31) → dt(0:31)
dt-a	dt(0:31) → ad(0:31)
dt-b	dt(0:31) → bd(0:31)

Figure 24. DT functions.

in the exception processing that can accompany many System/370 instructions. Columns 2 and 3 vary according to the needs of each
instruction.

Figure 26 lists the ALU input and output functions. The ALU
result register (ALUO) and several carry-out bits are latched whenever ALU calculates a new value. In addition to providing ALUO as
source data for the buses, ALU can generate a decimal arithmetic
correction factor (CORF) based on the carry out of each 4-bit nibble.

The Micro/370 ALU generates 7 condition signals. ALUC, ALUN,
ALUV, and ALUZ are the carry, negative, overflow, and zero flags for
the full 32-bit result. ALUBC, ALUBN, and ALUBZ are the carry,
negative, and zero flags for the low-order byte. The condition
signals are used by the Shifter, by the Branch Control, and by the
System/370 Condition Code Control. Every time ALU generates a new
result, Micro/370 updates the Condition Code (CC0 and CC1 — bits
18-19 of the PSW). IRE (containing the System/370 instruction) determines whether word or byte condition flags go into the formula
for calculating the new Condition Code. IRE and the ALU Column select the formula. Figure 27 shows the formulas.

2.14 DI

The 32-bit Data Input (DI) Register receives data from the real
(off-chip) world via Data Bus Control (DBC). Figure 28 lists its

ALU COLUMN (Selected by Nanoword)						INSTRUCTION(s)
1	2	3	4	5	6	(Selected by IRE)
A+B	A+B		A-B	AND	XOR	A,AH,AL,ALR,AR,ICM, LNR,LTR,SLA,SRA,TS
A+B		A-B	A-B	AND	XOR	BXH,CDS,CS
A+B		B-A	A-B	AND	XOR	BXLE
A+B	A-B		A-B	AND	XOR	C,CH,CL,CLC,CLCL,CLI, CLM,CLR,CR,LCR,LPR, MVCL,S,SH,SL,SLR,SR
A+B	A+B+1		A-B	AND	XOR	CVD
A+B	B-A		A-B	AND	XOR	D,DR,M,MH,MR
A+B	A+B	A+B+1	A-B	AND	XOR	DIAG4
A+B	A-B	A+B+1	A-B	AND	XOR	DIAG5
A+B		OR	A-B	AND	XOR	EX
A+B	AND		A-B	AND	XOR	N,NC,NI,NR,STNSM
A+B	OR		A-B	AND	XOR	O,OC,OI,OR,STOSM
A+B	A+B	A+B	A-B	AND	XOR	SLDA,SRDA
A+B	AND	XOR	A-B	AND	XOR	TM
A+B		XOR	A-B	AND	XOR	TRT
A+B	XOR		A-B	AND	XOR	X,XC,XI,XR
A+B			A-B	AND	XOR	All Others

Figure 25. ALU arithmetic and logic functions.

When the Flowcharts say:	Then:
a-alu	ad(0:31) → alua(0:31)
0-alu	0 → alua(0:31)
+1-alu	0 → alua(0:28); 001 → alua(29:31)
+2-alu	0 → alua(0:28); 010 → alua(29:31)
+3-alu	0 → alua(0:28); 011 → alua(29:31)
+4-alu	0 → alua(0:28); 100 → alua(29:31)
+6-alu	0 → alua(0:28); 110 → alua(29:31)
+7-alu	0 → alua(0:28); 111 → alua(29:31)
-1-alu	1 → alua(0:28); 111 → alua(29:31)
-2-alu	1 → alua(0:28); 110 → alua(29:31)
-3-alu	1 → alua(0:28); 101 → alua(29:31)
-4-alu	1 → alua(0:28); 100 → alua(29:31)
-6-alu	1 → alua(0:28); 010 → alua(29:31)
-7-alu	1 → alua(0:28); 001 → alua(29:31)
-8-alu	1 → alua(0:28); 000 → alua(29:31)
b-alu	bd(0:31) → alub(0:31)

Notes:
1. alua(0:31): ALU input A
   alub(0:31): ALU input B

When the Flowcharts say:	Then:
alu-a	aluo(0:31) → ad(0:31)
alu-b	aluo(0:31) → bd(0:31)
corf-a	1 → ad(0); aluc(0) → ad(1:2); 1 → ad(3:4); aluc(4) → ad(5:6); 1 → ad(7:8); aluc(8) → ad(9:10); 1 → ad(11:12); aluc(12) → ad(13:14); 1 → ad(15:16); aluc(16) → ad(17:18); 1 → ad(19:20); aluc(20) → ad(21:22); 1 → ad(23:24); aluc(24) → ad(25:26); 1 → ad(27:28); aluc(28) → ad(29:30); 1 → ad(31)

Notes:
1. aluo(0:31): ALU result register
   aluc(0:31): ALU carry-out register
2. 'corf' is a correction factor used in decimal arithmetic algorithms. Each nibble (4 bits) is 1001 or 1111, depending on the carry out of the corresponding nibble of the result.

Figure 26. ALU input functions (above) and output functions (below).

ALU COLUMN 1	2	3	4	5	6	DATA TYPE	INSTRUCTION(s) (Selected by IRE)
A	7		A	A	A	Word	A,AH,AR,LNR,LTR,SLA,SRA
A	D		A	A	A	Word	AL,ALR,SL,SLR
A		A	A	A	A	Word	BXH,BXLE,EX
A	1		A	A	A	Word	C,CH,CR
A		C	A	A	A	Word	CDS,CS
A	9		A	A	A	Word	CL,CLM,CLR
A	9		A	A	A	Byte	CLC,CLCL,CLI
A	A		A	A	A	Word	CVD,D,DR,M,MH,MR,STNSM,STOSM
A	D	E	A	A	A	Word	DIAG4,DIAG5
A	5		A	A	A	Word	ICM,MVCL
A	7		A	A	A	Word	LCR,LPR,S,SH,SR
A	C		A	A	A	Word	N,NR,O,OR,X,XR
A	C		A	A	A	Byte	NC,NI,OC,OI,XC,XI
A	7	6	A	A	A	Word	SLDA,SRDA
A	C	B	A	A	A	Byte	TM
A		8	A	A	A	Byte	TRT
A	4		A	A	A	Byte	TS
A			A	A	A		All Others

FORMULA	NEW CC0	NEW CC1
1	$\neg n \bullet \neg v \bullet \neg z + n \bullet v$	$n \bullet \neg v + \neg n \bullet v$
4	0	n
5	$\neg n \bullet \neg v \bullet \neg z$	n
6	$cc0 \bullet \neg n + cc1 \bullet \neg n + \neg n \bullet \neg z$	n
7	$\neg n \bullet \neg z + v$	$n + v$
8	$\neg z$	0
9	$c \bullet \neg z$	$\neg c$
A	$cc0$	$cc1$
B	$z \bullet cc1$	$cc1$
C	0	$\neg z$
D	c	$\neg z$
E	$cc0$	$\neg z$

Notes:
1. Blank column entries do not apply to those instructions.
2. ALU COLUMN and INSTRUCTION select FORMULA and DATA TYPE.
3. '•' = AND, '+' = OR, '¬' = NOT
4. ALU condition flags: 'c' = carry, 'n' = negative, 'v' = overflow, 'z' = zero
5. DATA TYPE selects Byte or Word flags for the formulas.
6. 'cc0' and 'cc1' are the previous S/370 condition code.

Figure 27.   System/370 Condition Code Control.

functions. DBC delivers 16 bits of data in one processor cycle. The data go to either the low-order half of DI or to both the low-order half and the high-order half of DI. As a data source for the internal E-Unit buses, DI can deliver several combinations of data widths, with leading zeroes or with sign extension.

When the Flowcharts say:	Then:
di-a	di(0:31) → ad(0:31)
di-b	di(0:31) → bd(0:31)
dibz-b	di(24:31) → bd(24:31); 0 → bd(0:23)
dile-b	di(16) → bd(0:15); di(16:31) → bd(16:31)
dilz-a	di(20:31) → ad(20:31); 0 → ad(0:19)
dilz-b	di(20:31) → bd(20:31); 0 → bd(0:19)
edb-di	dbc(0:15) → di(0:15); dbc(0:15) → di(16:31)
edb-dil	dbc(0:15) → di(16:31)
Notes: 1. 'dbc': data input from Data Bus Control	

Figure 28. DI functions.

2.15 DO

The 32-bit Data Output (DO) Register sends data to the real (off-chip) world via Data Bus Control (DBC). Figure 29 lists its functions. DO receives data from one of the internal E-Unit buses and, in the same processor cycle, passes either the high-order or the low-order 16 bits to DBC.

2.16 SFU

The 32-bit SFU Register links the components of the Special Function Unit with those in the main part of the E-Unit. Figure 30 lists the functions which can load data into SFU from either side. Components in the main part of the E-Unit transfer data to SFU only via the A bus. One form of transfer ('a-sfut') swaps two 4-bit nibbles. BSB, DIX, EXB, IDX, IRR, MIM, and MSB are located outside

When the Flowcharts say:	Then:
a-do	ad(0:31) → do(0:31)
b-do	bd(0:31) → do(0:31)
doh-edb	do(0:15) → dbc(0:15)
dol-edb	do(16:31) → dbc(0:15)
Notes: 1. 'dbc': data output to Data Bus Control	

Figure 29. DO functions.

When the Flowcharts say:	Then:
a-sfu and not bc-sfu	ad(0:31) → sfu(0:31)
a-sfu and bc-sfu	ilc(0:1) → sfu(0:1); pswh(18:23) → sfu(2:7); ad(8:31) → sfu(8:31)
a-sfut	ad(28:31) → sfu(24:27); ad(24:27) → sfu(28:31)
a1	if ird(2:3) = 11: ird(12:15) → sfu(28:31); 0 → sfu(0:27) if ird = EX instruction: ird(0:15) → sfu(16:31); 0 → sfu(0:15) otherwise: ird(8:15) → sfu(24:31); 0 → sfu(0:23)
bsb-sfu	bsb(4:6) → sfu(28:30); 0 → sfu(0:27); 0 → sfu(31)
exb-sfu	after di blowup: dix(0:4) → sfu(24:28); 0 → sfu(0:23); 0 → sfu(29:31) after ird blowup: idx(0:4) → sfu(24:28); 0 → sfu(0:23); 010 → sfu(29:31) otherwise: exb(0:7) → sfu(24:31); 0 → sfu(0:23)
ilc-sfu	ilc(0:1) → sfu(29:30); 0 → sfu(0:28); 0 → sfu(31)
ipb-sfu	irr(1:4) → sfu(24:27); irr(5:7) → sfu(29:31); 0 → sfu(0:23); 0 → sfu(28)
ire-sfu	ire(0:15) → sfu(16:31); 0 → sfu(0:15)
lvl-sfu	irr(5:7) → sfu(29:31); 0 → sfu(0:28)
ll-sfu	ire(8:11) → sfu(28:31); 0 → sfu(0:27)
mim-sfu	pswh(6) → sfu(24); mim(1:7) → sfu(25:31); 0 → sfu(0:23)
msb-sfu	mode → sfu(27); ors → sfu(28); ows → sfu(29); wrs → sfu(30); rrs → sfu(31); 0 → sfu(0:26)
pswh-sfu	pswh(0:31) → sfu(0:31)

Notes:
1. 'a1': start executing either a new S/370 instruction or the target instruction of an EX instruction.
2. 'di blowup': an operand displacement is needed, but the di register has invalid data because of a Bus Access error.
3. 'ird blowup': a new instruction must be decoded, but the ird register has invalid data because of a Bus Access error.

Figure 30. SFU input functions.

the Special Function Unit and transfer data to SFU via BSS, an 8-bit port on the low-order byte of the C bus.

Figure 31 lists the functions in which SFU sends data either to the main part of the E-Unit (only via the B bus) or to components of the Special Function Unit. BSB and MIM receive data from SFU via the BSS port.

32

When the Flowcharts say:	Then:
sfu-b	sfu(0:31) → bd(0:31)
sfu-bsb	sfu(28:30) → bsb(4:6)
sfu-cc	sfu(30:31) → pswh(18:19)
sfu-ilc	sfu(29:30) → ilc(0:1)
sfu-ird	sfu(16:31) → ird(0:15)
sfu-ire	sfu(16:31) → ire(0:15)
sfu-mim	sfu(25:31) → mim(1:7)
sfu-pswh	sfu(0:31) → pswh(0:31)

Figure 31.   SFU output functions.

When Micro/370 starts processing a System/370 instruction, SFU contains one or more 4-bit nibbles of the instruction (transferred from IRD), depending on the instruction.   (See 'a1' in Figure 30.)

2.17   IRF, IRD, and IRE

Micro/370 uses three halfword (16-bit) registers to prefetch instructions.   They are part of the Special Function Unit. Figure 32 lists their functions. When Micro/370 starts processing a System/370 instruction, IRD and IRE each contain the first halfword of the instruction, and IRF contains the following halfword.

When the Flowcharts say:	Then:
a1	ird(0:15) → ire(0:15)
edb-irf	dbc(0:15) → irf(0:15)
ird-ire	ird(0:15) → ire(0:15)
irf-ird	irf(0:15) → ird(0:15)
inc r1	{ire(8:11) + 1} → ire(8:11)

Notes:
1. 'a1': start executing either a new S/370 instruction or the target of an EX instruction.
2. 'dbc': data output to Data Bus Control

Figure 32.   IRF, IRD, and IRE functions.

2.18 ILC

ILC, another component of the Special Function Unit, is the 2-bit System/370 instruction-length code. It gives the length of the current System/370 instruction in halfwords. When Micro/370 starts processing a System/370 instruction, ILC receives a new value, based on the instruction in IRD. Figure 30 and Figure 31 list, among others, the functions which can transfer data between ILC and SFU.

2.19 PSWH

Micro/370 maintains the high-order half of the System/370 Program Status Word in PSWH, a 32-bit register in the Special Function Unit. Figure 6 shows the required EC-Mode format. Figure 30 and Figure 31 list, among others, the functions which can transfer data between PSWH and SFU. In addition, the two bits PSWH(18:19) receive a new System/370 Condition Code whenever Micro/370 uses ALU. (See Figure 27.)

3 STATE MACHINE CONTROL

The Micro/370 Processor is controlled by a State Machine, with its states represented in microcode. I described the Control Store in section "1.3 The Micro/370 Chip Floorplan". State Machine Control selects the next Control Store Address (CSA), thereby selecting the next state. In this section I list the various mechanisms used for that selection. Figure 33 shows the components of State Machine Control.

3.1 Direct Branches

Most states select one specific next state. The Microword contains the full 10 bits of the next state's CSA.

3.2 Conditional Branches

The flowcharts use conditional branches to make data-dependent control decisions. The Microword uses 5 bits to select one of 31 conditional branch types. It also contains 7 bits of the next state's CSA. The high-order bit of the CSA comes from the high-order bit (inverted) of the branch type. The Branch Control PLA generates the 2 low-order bits of the CSA according to a formula selected by the branch type. Various condition signals from the E-Unit form the inputs to those formulas. Figure 34 lists the formulas.

3.3 Instruction Decoders

The flowcharts use 'ad1' or 'ad2' to tell (via the Microword) State Machine Control to select the next CSA from the A1 or A2 Instruction Decode PLA. Many System/370 instructions having a similar format require similar operations, such as fetching one or two operands or prefetching the following instruction. The Micro/370 Flowcharts take advantage of the similarity by using a common microsequence, selected by the A1 decoder. At the end of the common microsequence, the flowcharts use the A2 decoder to branch to individual microsequences dependent on the System/370 instruction. To execute similar operations at the end of processing, the flowcharts branch directly to common exit states, which finally conclude by selecting the A1 decoder, with IRD containing the next instruction.

The decoders take their input from IRD, which holds the System/370 instruction, and certain mode flags, which enable certain instructions. The selected CSA corresponds to the first state of the microsequence appropriate for processing that instruction. If the mode flags have disabled an instruction, then the decoder points to a microsequence which generates a Privilege Exception or an Operation Exception. If the instruction in IRD is not in the set of instructions which Micro/370 implements, the decoder selects the CSA for an Operation Exception.

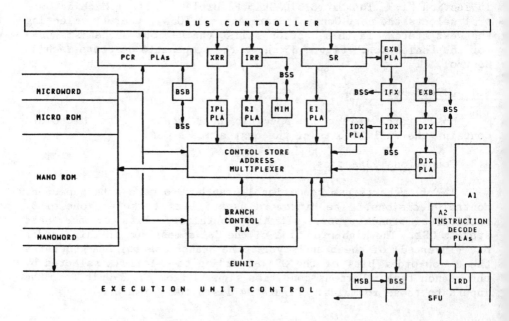

Figure 33. Micro/370 State Machine Control.

bc00	cc=true cc=false	bc12	(alu>=0)(au>=0) (alu<0)+(au<0)	bc22	sh(0)=0 sh(0)=1
bc01	(alu>=0)+(au>=0) (alu<0)(au<0)	bc13	au<0 au>=0	bc23	(sh(0)=0)(au¬=0) (sh(0)=1)(au¬=0) au=0
bc02	alu<0 alu>=0	bc14	alu=0 alu¬=0	bc24	aluc=1 aluc=0
bc03	aluv=0 aluv=1	bc15	(alu=0)(au<0) (alu=0)(au>=0) (alu¬=0)(au<0) (alu¬=0)(au>=0)	bc25	ire(12:13)=0 ire(12:13)¬=0
bc04	pswh(20)=0 pswh(20)=1	bc16	sl(31)=0 sl(31)=1	bc26	ire(14:15)=00 ire(14:15)=01 ire(14:15)=10 ire(14:15)=11
bc05	alu>0 alu<=0	bc17	r1=r3 r1¬=r3	bc27	ire(12:13)=00 ire(12:13)=01 ire(12:13)=10 ire(12:13)=11
bc06	alu<0 alu=0 alu>0	bc18	digit ¬digit		
bc07	(alu=0)(au=0) (alu¬=0)+(au¬=0)	bc19	(pos)(alu<0) (pos)(alu>=0) neg digit	bc28	sl(30:31)=0 sl(30:31)¬=0
bc08	ovf=0 ovf=1	bc20	alub=0 alub¬=0	bc29	au(24:31)=0 au(24:31)¬=0
bc09	(alu<0)(au<0) (alu<0)(au>=0) (alu>=0)(au<0) (alu>=0)(au>=0)	bc21	au=0 au¬=0	bc30	auc=1 auc=0
bc10	pswh(14)=0 pswh(14)=1	bc31	au(24:31)=0 (au(24:31)¬=0)(sl(30:31)=01) (au(24:31)¬=0)(sl(30:31)=10) (au(24:31)¬=0)(sl(30:31)=00,11)		

Notes:
1. cc: 370 condition code (0, 1, 2, or 3)
   cc=true: (cc=0 AND ire(8)=1) OR (cc=1 AND ire(9)=1) OR
            (cc=2 AND ire(10)=1) OR (cc=3 AND ire(11)=1)
2. ovf: arithmetic left shift overflow
3. r1=r3: ire(8:11) = ire(12:15)
4. digit: sfu(28:31) = decimal digit (0000-1001)
5. pos: sfu(28:31) = decimal plus sign (1010,1100,1110,1111)
6. neg: sfu(28:31) = decimal minus sign (1011,1101)
7. alub: alu(24:31)

Figure 34.   Conditional Branches.

3.4   Bus Access Exceptions

The Micro/370 Processor requests external Bus Access operations via the Bus Controller's Processor Command Register (PCR). Decoders translate 5 bits of the Microword into the PCR format. For some Service Cycle requests, the 3-bit BSB register specifies 3 bits of the PCR. When the Bus Controller completes its operations, it signals State Machine Control via its Status Register (SR). With normal completion, this has no effect on the CSA selection. Abnormal terminations, however, can affect the processing either immediately or with some delay, depending on the type of Bus Access and on the particular form of error detected by the Bus Controller.

The EXB register records an encoding of the completion status after each Bus Access. If the Bus Access was supposed to read a halfword of the instruction stream into IRF, IFX also records the encoded status. If the Bus Access was supposed to read a halfword of the instruction stream into DI, the encoded status recorded in EXB is passed to DIX. When IRD receives the halfword in IRF, in some later processor cycle, IDX also receives the corresponding status in IFX. That is, IFX, IDX, and DIX always contain an encoded record of the Bus Controller's status signal after it fetched the contents of IRF, IRD, and DI, respectively.

When the output of either Instruction Decoder is selected for the next CSA, but the halfword in IRD is invalid because of an error when it was originally read into IRF, the encoded status in IDX selects the CSA for an exception-handling microsequence. The CSA Multiplexer notes the presence of this alternative CSA at the time when it would otherwise select the Instruction Decoder's output. The Multiplexer chooses the CSA from the IDX PLA, and Micro/370 generates an appropriate interruption.

Likewise, the CSA Multiplexer recognizes when the E-Unit is using the contents of DI to calculate an operand address from a base register and a displacement. When DIX signals that the halfword in the lower half of DI is invalid, the Multiplexer chooses the CSA from the DIX PLA, and Micro/370 generates an appropriate interruption.

In both cases, Micro/370 does not signal the Bus Access error in prefetching instructions until it actually requires the instruction. This is a requirement of System/370 architecture. Errors in reading or writing operands, however, must be processed immediately. The EI (Exigent Interruption) PLA in this case takes priority over the selection of the next CSA, pointing to a microsequence appropriate to the error indicated by SR.

3.5 External Interrupts

The Bus Controller synchronizes external interrupt signals and passes them to State Machine Control via its Interrupt Request Register (IRR). This mechanism can interrupt normal processing only between System/370 instructions, that is, only when the A1 Instruction Decoder would otherwise generate the next CSA. The interrupts include the 5 System/370 Interruptions (External, I/O, Machine Check, Program, and System/370 Restart), System/370 Start and System/370 Stop signals, and the 7 M68000 interrupts. In Motorola's M68000 architecture, its 7 interrupt levels are ranked in priority and disabled collectively by raising or lowering a priority level in the processor's Status Register. In Micro/370, a DIAGNOSE instruction enables each of the 7 interrupt levels independently by setting the 7-bit Motorola Interrupt Mask (MIM). The RI (Repressible Interrupt) PLA detects the external interrupt signals of IRR and, subject to masking by MIM, generates the CSA appropriate for the microsequence that handles the interrupt.

3.6 External Resets

The Bus Controller synchronizes external reset signals and passes them to State Machine Control via its External Reset Register (XRR). There are 3 types of external resets: CPU Reset, Initial CPU Reset, and Initial Program Load (IPL). The IPL PLA detects the external reset signal of XRR and generates the CSA appropriate for the microsequence that handles the reset. Micro/370 responds to resets according to System/370 architecture. It uses a Service Cycle to obtain the IPL PSW when required.

4 PROCESSOR TIMING

A 4-phase 6-output clock phase generator times Micro/370 Processor operations. The generator is driven by an external (off-chip) crystal oscillator to produce phase signals which each last approximately half of a clock cycle. During a processor cycle, the phase generator produces four phases, named 'P1', 'P2', 'P3', and 'P4'. P1 and P3 occur when the clock is high; P2 and P4 occur when the clock is low. The normal processor cycle takes 2 clock cycles (2T, where T = one clock cycle). The Processor can extend a processor cycle whenever it must wait for the Bus Controller to complete a Bus Access. In this case, the clock phase generator extends the normal processor cycle through additional clock cycles, until the Bus Controller passes its completion status to State Machine Control. The additional clock cycles follow P4 and consist of no signal during the first half (clock high) and another P4 during the second half (clock low). Once the Bus Access is complete, the phase generator proceeds at the start of the next clock cycle to P1, which begins the next processor cycle.

The clock phase generator produces two additional phase signals. 'PP' overlaps P1 and P2 and controls the precharging of dynamic circuits in the E-Unit. 'PR' occurs at the same time as the first P4 signal in each processor cycle. Figure 35 lists the purposes of PR as well as P1 - P4.

During Phase:	This happens in the E-Unit:	And this happens in State Machine Control:
p1	source data → buses IRF → IRD S/370 CC → PSWH control signals latch	Multiplexer latches new CSA Branch Control gets condition signals from E-Unit
p2	buses transfer data AUO → AO AO → ABC	Microrom fetches Microword Nanorom fetches Nanoword
p3	buses → destinations ALUO → DT AUO → PC, PSWL DO → DBC IRD → IRE	Microword and Nanoword latch EXB → DIX IFX → IDX
pr	ALU, AU, Shifter latch results PLAs generate controls	PLAs prepare new PCR for Bus Controller
p4	DBC → DI, IRF buses precharge	PLAs get new SR from Bus Controller EXB and IFX latch

Figure 35. Processor functions during each clock phase.

5 BUS CONTROLLER

The Micro/370 Bus Controller is in charge of the chip's 106 signal pins. It performs four functions:

- responds to the Processor's requests for Bus Access

- notifies the Processor of External Interrupts

- notifies the Processor of External Resets

- arbitrates use of the Micro/370 Bus

5.1 Micro/370 Pins

Note: The bit numbering of Micro/370 pins is the reverse of IBM bit numbering (noted in section "1.2 What Micro/370 Does"). The least-significant pin has the smallest number, in agreement with the notation of Motorola's M68000 component bus.

5.1.1 Address Bus.

- A31-A01 provide 31 bits of the address for bus operations. They address a halfword of data. A31 is the most-significant bit of a 32-bit byte address. The least-significant bit is omitted, because the data strobes select the upper or lower byte for byte transfers. During a Motorola Interrupt Acknowledge Cycle, bits A03-A01 are the interrupt level. During Micro/370 Service Cycles, bits A03-A01 are all 0, and bits A31-A04 carry addressing information.

- A32 is a special signal that identifies the address as Real (A32 = 1) or Virtual (A32 = 0).

5.1.2 Data Bus. D31-D00 are the 32-bit bidirectional data path for bus operations. Data transfers can have data widths of 1 byte, 2 bytes (halfword), or 4 bytes (word).

5.1.3 Function Code. FC2-FC0 are 3 bits which identify the address space (type of memory or device selected) for bus operations. Function Code 7 identifies a Motorola Interrupt Acknowledge Cycle (if A03-A01 ¬= 0) or a Service Cycle (if A03-A01 = 0).

5.1.4 Bus Cycle Controls.

- Read/Write (R/¬W): identifies the direction of data transfer during a bus cycle. R/¬W = 1 means that Micro/370 is reading data, transferring data from a device to the processor.

- Address Strobe (AS): signifies valid information on the address bus.

- Upper Data Strobe (UDS): for a Read Bus Cycle — requests the device to provide on D31-D24 the upper byte of the addressed halfword. For a Write Bus Cycle — UDS signifies valid data on D31-D24, which the device should store in the upper byte of the addressed halfword.

- Lower Data Strobe (LDS): for a Read Bus Cycle — requests the device to provide on D23-D16 the lower byte of the addressed halfword. For a Write Bus Cycle — LDS signifies valid data on D23-D16, which the device should store in the lower byte of the addressed halfword.

- Word Strobe (WS): for a Read Bus Cycle — does not apply. For a Write Bus Cycle — WS signifies valid data on D31-D00, which the device should store in the addressed word. Micro/370 asserts WS only when A01 = 0, so that the data are aligned on a word boundary.

- Data Transfer Acknowledge (DTACK): tells the Bus Controller that the device has transferred the byte or bytes signified by UDS and LDS. If the Bus Controller was trying to write a full word (with WS), only the upper halfword was transferred; the Bus Controller will run a second Write Bus Cycle to transfer the lower halfword on D31-D16.

- Word Transfer Acknowledge (WTACK): for a Read Bus Cycle — tells the Bus Controller that the device has placed on D31-D00 the full word of data addressed by A31-A02. Data Bus Control must select the requested byte(s) of data to pass to the E-Unit. If the Processor had requested a word of data, and the data were word-aligned (A01 = 0), then the Bus Controller saves the lower halfword in a buffer, while Data Bus Control passes the upper halfword to the E-Unit. For a Write Bus Cycle — WTACK tells the Bus Controller that the device transferred all of the data.

- Byte Transfer Acknowledge (BTACK): for a Read Bus Cycle — tells the Bus Controller that the device has placed on D31-D24 the first byte requested. If the Bus Controller was trying to read a halfword (with UDS and LDS), only the upper byte was transferred; the Bus Controller will run a second Read Bus Cycle to transfer the lower byte. For a Write Bus Cycle — BTACK tells the Bus Controller that the device transferred only the first byte of data. If the Bus Controller was trying to write more than a byte, it will run more bus cycles to transfer the remaining data.

- Bus Error (BERR): tells the Bus Controller to terminate the bus cycle without transferring any data. In combination with HALT, it can tell the Bus Controller to re-run the bus cycle.

- Data Transfer Exception (DTEXN/2-DTEXN/0): tell the Bus Controller to terminate the bus cycle without transferring any data. The Bus Controller reports to the Processor the specific exception identified by the 3 signals.

5.1.5 Interrupt Controls.

- Interrupt Priority Level (IPL2-IPL0): Motorola Interrupt. The 3 bits signify the Interrupt Level.

- Priority Request (PR1-PR0): 0 = undefined; 1 = System/370 Stop; 2 = System/370 Start; 3 = E-Unit Dump.

- System/370 Interruption Requests (IR4-IR0): IR4 = Restart; IR3 = External; IR2 = Program; IR1 = Machine Check; IR0 = I/O.

5.1.6 Bus Arbitration Controls. Bus Request (BR), Bus Grant (BG), and Bus Grant Acknowledge (BGACK) perform the same functions as in the Motorola M68000 architecture.

5.1.7 M6800 Peripheral Controls. Valid Peripheral Address (VPA), Valid Memory Address (VMA), and Enable (E) perform the same functions as in the Motorola M68000 architecture. During bus cycles, the Bus Controller's response to VPA is similar to its response to BTACK.

5.1.8 Reset Controls.

• Auxiliary Reset (XRESET): modifies External Reset requests;
  tells the Bus Controller to ignore an erroneous device response
  during a Bus Cycle.

• Reset (RESET): bidirectional signal. A DIAGNOSE instruction can
  tell Micro/370 to drive the signal in order to reset attached
  devices.

An external device can drive the Reset Controls in combination with
HALT to generate an External Reset signal to the Processor: RESET
and HALT = Initial CPU Reset; XRESET and HALT = CPU Reset; RESET,
XRESET, and HALT = IPL.

5.1.9 System/370 State Indicators.   I/1-I/0 indicate the current
state of the Micro/370 Processor: 0 = Load or Operating; 1 = Stopped;
2 = Checkstop; 3 = undefined.

5.1.10 Timing.

• Input Clock (CLK): the external clock drives all operations on
  the Micro/370 chip. The chip runs nominally at 20 MHz and is
  guaranteed to function at 10 MHz.

• Bus Cycle in Progress (BCIP): signifies the start of a bus cycle.
  It gives advance warning, one clock cycle earlier than AS.

5.1.11 Miscellaneous Controls.

• Select Read Cycle (3T/¬4T): enables the Bus Controller to attempt to complete a Read Bus Cycle in 3 clock cycles (3T) by
  looking for a device response 1T earlier than required by the
  Motorola M68000 bus protocol.

• SUPPRESS: prevents the Bus Controller from attempting to write
  a full word of data in a single Write Bus Cycle.

• DUAL: enables the Micro/370 Dual Mode Feature. Unless this pin
  is asserted, Micro/370 will always operate in 370 Mode, and all
  bus cycles will access 370 Space.

• Macrocycle in Progress (MIP): signifies that Micro/370 is running a series of bus cycles which can not be interrupted. Bus
  arbitration is suspended. Micro/370 asserts MIP when it must
  perform a Read-Modify-Write or when it swaps PSWs for an interruption.

• HALT: a bidirectional signal. Micro/370 asserts HALT only when
  it enters the Checkstop state. If an external device asserts

HALT alone, it has no effect. If the Bus Controller detects HALT at the same time as a normal acknowledge to a bus cycle, it will not start another bus cycle until HALT is removed. HALT also serves as a modifier to BERR and must be asserted with the Reset Controls to cause an External Interrupt.

5.2 Bus Access

The Micro/370 Flowcharts include terms to describe 29 types of Bus Access requests. A set of PLAs translate 5 bits of the Microword into the format of the Bus Controller's Processor Command Register (PCR). The Bus Controller may satisfy the request with one or more Bus Cycles, or it may be able to satisfy it with an internal Buffer Cycle, which is not seen by the real world. In this section I discuss the circumstances which dictate how the Bus Controller responds to the Processor's Bus Access requests.

Bus Controller operations depend on the alignment of the data. If the Processor specifies 'W' (word) for the access width, the Processor transfers the upper halfword in this operation and will transfer the lower halfword in its next Bus Access request. The Bus Controller, however, treats a word request the same as a halfword ('H') request, unless the address for the upper halfword is on a full word boundary.

Note: Figure 36 and Figure 37 contain all of the notes pertaining to the figures I show in this section. Refer to it when in doubt about the names and abbreviations I use.

5.2.1 Address Spaces. The Bus Access terms in the flowcharts correspond to the address space which the Processor wishes to access. Figure 38 shows the correspondence. The first column of the table is the Bus Access in Micro/370 Flowchart notation. The second column contains some additional factors which affect the selection of address space. 'm = 1' when Dual Mode operation has been enabled (by the 'DUAL' control pin) and Micro/370 is operating in Control Mode (indicated by the Processor's 'MODE' control flag). 'r = 1' when 'm = 1' and, additionally, a DIAGNOSE instruction has told Micro/370 to read operands from Control Space (indicated by the Processor's 'ORS' control flag). 'w = 1' when 'm = 1' and, additionally, a DIAGNOSE instruction has told Micro/370 to write operands to Control Space (indicated by the Processor's 'ORS' control flag). 'a = 0' when the 3 bits AO(28:30) are all 0. These are the 3 bits which Address Bus Control (ABC) passes to Address Bus lines A03-A01. With Function Code 7, Motorola M68000 architecture interprets a Read Cycle as an Interrupt Acknowledge Cycle, expecting the Interrupt Level (a number from 1 to 7) to appear on lines A03-A01. Micro/370 extends the use of Function Code 7 to Sense (Input) Service Cycles and Con-

Titles:
   AO(30:31): the two least-significant bits of AO(0:31), which is the 32-bit byte address.
   SUPPRESS: a pin which, when active (1), suppresses full-word write cycles.

When the Flowcharts say:
   A state requesting bus access 'irin' (or 'irop') MUST be followed immediately by a state
   with access 'frin' (or 'frop', respectively). 'frin' ('frop') has no effect on the Bus
   Controller. The time required by the Bus Controller to respond to the 'irin' ('irop')
   command includes the 'frin' ('frop') state.

And:
   'm' = dual & mode = Control Mode
   'r' = dual & mode & ors = read operands from Control Space
   'w' = dual & mode & ows = write operands to Control Space
   'dual': a pin which enables Dual Mode operation.
   'mode', 'ors', and 'ows': flags controlled by the Processor.
   'a' = AO(28:30), which the Bus Controller passes to address pins A3, A2, and A1 during a
   Service Cycle. They are all zero for Sense and Control Service Cycles. They are the
   non-zero Interrupt Level for an Interrupt Acknowledge Cycle.

Then the PCR says:
   Sup: 1 = Control (Supervisor) Space; 0 = 370 (User) Space
   Prg: 1 = Instruction (Program) Space; 0 = Operand (Data) Space
   Cyc or Cycle: 001 = Read Cycle; 010 = Write Cycle;
                 101 = Input Service Cycle; 110 = Output Service Cycle
   Width: 001 = Byte; 010 = Halfword (16 bits); 011 = Word
   StSel: Strobe Selection for Service Cycles. For srmc and swmc accesses, the 3-bit 'bsb'
   (Bus Strobe Bits) register is set by the Processor to produce an appropriate Sense or
   Control Service Cycle. (For example, bsb = '011' produces Sense/3 and Control/3 Service
   Cycles, which communicate with a coprocessor.) For srpg and swpg accesses, the 'bsb'
   register is loaded from the second operand address of a DIAGNOSE instruction.
   I: 1 = Inhibit 3T Read. This prevents the Bus Controller from returning status to the
   Processor before the Processor is ready to accept it. It is needed when the Processor is
   executing two states (irin,frin or irop,frop), which require 4T.

Figure 36.   Bus Access Notes (Part 1).

44

If WIP is:

WIP (Word In Progress) is a flag set and reset by the Bus Controller to indicate that it has buffered internally a halfword of data as part of its response to a Read Word or Write Word command (PCR Cycle 001 or 010, Width 011).

If SWIP is:

SWIP (Service Word In Progress) is a flag set and reset by the Bus Controller to indicate that it has buffered internally a halfword of data as part of its response to a Service Cycle command (PCR Cycle 101 or 110).

Then the Bus Controller does:

ICU (Interface Control Unit): the Micro/370 Bus Controller

When the Bus Controller must perform more than one bus cycle in response to a single Processor command, subsequent cycles are numbered 2, 3, 4. This occurs (a) when the memory or device indicates by its acknowledge signal (DTACK or BTACK) that it was not able to accept all of the data sent to it, or (b) when the Processor is accessing a halfword of data at an odd byte address.

Except for timing differences, a VPA acknowledge signal results in the same Bus Controller action as BTACK.

And the Outside World sees:

A32: 0 = virtual address; 1 = real address

FC: 1 = 370 (User) Operand (Data) Space   2 = 370 Instruction Space
    5 = Control (Supervisor) Operand Space   6 = Control Instruction Space
    7 = Service Cycle or Interrupt Acknowledge Cycle

R/¬W: 1 = Read; 0 = Write   WS: 1 = Word Strobe active

UDS: 1 = Upper Data Strobe active   LDS: 1 = Lower Data Strobe active

And it takes at least:

T = 1 clock cycle.

Minimum time shown is for the fastest device response. Times for Read or Write Cycles for slower devices must be adjusted to whatever time they require to complete a bus cycle.

For example, if a memory can meet the timing requirements for a 4T Read Cycle but not for a 3T Read, then all Read Cycle times should be calculated with 4T, including those already shown in the tables as 4T.

Times required to access M6800 devices are much longer.

Figure 37. Bus Access Notes (Part 2).

When the Flow-charts say:	And:	Then the PCR says:			And the Outside World sees:			Which means:
Access		Sup	Prg	Cyc	A32	FC	R/¬W	Bus Cycle
irin	m = 0	0	1	001	0	2	1	Fetch Instruction from Virtual 370 Space
	m = 1	1	1	001	0	6	1	Fetch Instruction from Virtual Control Space
irop	r = 0	0	0	001	0	1	1	Fetch Operand from Virtual 370 Space
	r = 1	1	0	001	0	5	1	Fetch Operand from Virtual Control Space
trin	m = 0	0	1	001	0	2	1	Fetch Instruction from Virtual 370 Space
	m = 1	1	1	001	0	6	1	Fetch Instruction from Virtual Control Space
trop	r = 0	0	0	001	0	1	1	Fetch Operand from Virtual 370 Space
	r = 1	1	0	001	0	5	1	Fetch Operand from Virtual Control Space
trow	w = 0	0	0	001	0	1	1	Fetch Operand from Virtual 370 Space
	w = 1	1	0	001	0	5	1	Fetch Operand from Virtual Control Space
trro	r = 0	0	0	001	1	1	1	Fetch Operand from Real 370 Space
	r = 1	1	0	001	1	5	1	Fetch Operand from Real Control Space
wrop	w = 0	0	0	010	0	1	0	Store Operand into Virtual 370 Space
	w = 1	1	0	010	0	5	0	Store Operand into Virtual Control Space
wrro	w = 0	0	0	010	1	1	0	Store Operand into Real 370 Space
	w = 1	1	0	010	1	5	0	Store Operand into Real Control Space
srak	a = 0			101	1	7	1	Sense Service Cycle
	a ¬ 0			101	1	7	1	Level a Interrupt Acknowledge
srmc srpg	a = 0			101	1	7	1	Sense Service Cycle
swak swmc swpg	a = 0			110	1	7	0	Control Service Cycle

Figure 38. Bus Access Address Spaces.

trol (Output) Service Cycles, distinguishing them from Interrupt
Acknowledge Cycles by requiring A03-A01 to be 0.

The third column of the table of address spaces shows how the
PCR reflects the Bus Access type. The fourth column shows what appears on the 5 address space selection pins if a bus cycle takes
place. The last column explains what those 5 signals mean to the
real world.

5.2.2 AO(30:31) = 00.   When the 2 lowest bits of the byte address,
AO(30:31), are both 0, the data are aligned on a word boundary.
Figure 39 shows what the Bus Controller does in this case when SUP-
PRESS is not active. During Read Cycles, if the Processor requested
a full word and the device responds with WTACK, the Bus Controller
saves the lower halfword in a buffer and sets the Word in Progress
(WIP) flag. When the Processor requests a full word write, the Bus

When the Flow- charts say:	Then the PCR says:			Then the Bus Controller does:	And the Outside World sees:			And it takes at least:
Access	Cycle Width		I	ICU Sequence	WS	UDS	LDS	
irin,w frin irop,w frop	001	011	1	Read Cycle if WTACK - set WIP if BTACK - Read Cycle 2	0 0	1 0	1 1	4T 3T
trin,w trop,w trro,w	001	011	0	Read Cycle if WTACK - set WIP if BTACK - Read Cycle 2	0 0	1 0	1 1	3T 3T
irin,h frin irop,h frop	001	010	1	Read Cycle if BTACK - Read Cycle 2	0 0	1 0	1 1	4T 3T
trin,h trop,h trow,h trro,h	001	010	0	Read Cycle if BTACK - Read Cycle 2	0 0	1 0	1 1	3T 3T
irop,b frop	001	001	1	Read Cycle	0	1	0	4T
trop,b trow,b	001	001	0	Read Cycle	0	1	0	3T
wrop,w wrro,w	010	011	0	Buffer Write, set WIP				3T
wrop,h wrro,h	010	010	0	Write Cycle - if BTACK - Write Cycle 2	0 0	1 0	1 1	4T 4T
wrop,b	010	001	0	Write Cycle	0	1	0	4T

Figure 39. Bus Access with AO(30:31) = 00 and SUPPRESS = 0.

Controller saves the halfword in a buffer and sets the WIP flag.
The last column of the table shows, in clock cycles, how long it
takes (minimum) to complete the stated operation. The Buffer Write
requires 3T to determine that the data are word-aligned.

Figure 40 shows what happens in similar cases but with SUPPRESS
active. The Buffer Write is inhibited, so that a word write request
is treated the same as a halfword write.

5.2.3 AO(30:31) = 10. When the 2 lowest bits of the byte address,
AO(30:31), are 10, the data are aligned on a word boundary plus 2.
Figure 41 shows what the Bus Controller does in this case when SUP-
PRESS is not active. When the Processor requests a halfword read,
and WIP is set, then the Bus Controller must have saved the requested
halfword in its buffer. It performs a Buffer Read to give the
Processor its data. When the Processor requests a halfword write,

When the Flow-charts say:	Then the PCR says:			Then the Bus Controller does:	And the Outside World sees:			And it takes at least:
Access	Cycle	Width	I	ICU Sequence	WS	UDS	LDS	
irin,w frin irop,w frop	001	011	1	Read Cycle if WTACK - set WIP if BTACK - Read Cycle 2	0 0	1 0	1 1	4T 3T
trin,w trop,w trro,w	001	011	0	Read Cycle if WTACK - set WIP if BTACK - Read Cycle 2	0 0	1 0	1 1	3T 3T
irin,h frin irop,h frop	001	010	1	Read Cycle if BTACK - Read Cycle 2	0 0	1 0	1 1	4T 3T
trin,h trop,h trow,h trro,h	001	010	0	Read Cycle if BTACK - Read Cycle 2	0 0	1 0	1 1	3T 3T
irop,b frop	001	001	1	Read Cycle	0	1	0	4T
trop,b trow,b	001	001	0	Read Cycle	0	1	0	3T
wrop,w wrro,w	010	011	0	Write Cycle - if BTACK - Write Cycle 2	0 0	1 0	1 1	4T 4T
wrop,h wrro,h	010	010	0	Write Cycle - if BTACK - Write Cycle 2	0 0	1 0	1 1	4T 4T
wrop,b	010	001	0	Write Cycle	0	1	0	4T

Figure 40. Bus Access with AO(30:31) = 00 and SUPPRESS = 1.

Figure 41 (page 48)

When the Flowcharts say:	Then the PCR says:			If WIP is:	Then the Bus Controller does:	And the Outside World sees:			And it takes at least:
Access	Cycle	Width	I		ICU Sequence	WS	UDS	LDS	
irin,w frin irop,w frop	001	011	1		Read Cycle if BTACK - Read Cycle 2	0 0	1 0	1 1	4T 3T
trin,w trop,w trro,w	001	011	0		Read Cycle if BTACK - Read Cycle 2	0 0	1 0	1 1	3T 3T
irin,h frin irop,h frop	001	010	1	0	Read Cycle if BTACK - Read Cycle 2	0 0	1 0	1 1	4T 3T
trin,h trop,h trow,h trro,h	001	010	0	0	Read Cycle if BTACK - Read Cycle 2	0 0	1 0	1 1	3T 3T
irin,h frin irop,h frop	001	010	1	1	Buffer Read, reset WIP				4T
trin,h trop,h trro,h	001	010	0	1	Buffer Read, reset WIP				2T
irop,b frop	001	001	1		Read Cycle	0	1	0	4T
trop,b trow,b	001	001	0		Read Cycle	0	1	0	3T
wrop,w wrro,w	010	011	0		Write Cycle - if BTACK - Write Cycle 2	0 0	1 0	1 1	6T 4T
wrop,h wrro,h	010	010	0	0	Write Cycle - if BTACK - Write Cycle 2	0 0	1 0	1 1	4T 4T
wrop,h wrro,h	010	010	0	1	Write Cycle, reset WIP - if DTACK - Write Cycle 2 - then if BTACK - Cycle 3 - if BTACK - Write Cycle 2 and Write Cycle 3 - then if BTACK - Cycle 4	1 0 0 0 0 0	1 1 0 0 1 0	1 1 1 1 1 1	4T 4T 4T 4T 4T 4T
wrop,b	010	001	0		Write Cycle	0	1	0	4T

Figure 41. Bus Access with AO(30:31) = 10 and SUPPRESS = 0.

When the Flow-charts say:	Then the PCR says:			If WIP is:	Then the Bus Controller does:	And the Outside World sees:			And it takes at least:
Access	Cycle	Width	I		ICU Sequence	WS	UDS	LDS	
irin,w frin irop,w frop	001	011	1		Read Cycle if BTACK - Read Cycle 2	0 0	1 0	1 1	4T 3T
trin,w trop,w trro,w	001	011	0		Read Cycle if BTACK - Read Cycle 2	0 0	1 0	1 1	3T 3T
irin,h frin irop,h frop	001	010	1	0	Read Cycle if BTACK - Read Cycle 2	0 0	1 0	1 1	4T 3T
trin,h trop,h trow,h trro,h	001	010	0	0	Read Cycle if BTACK - Read Cycle 2	0 0	1 0	1 1	3T 3T
irin,h frin irop,h frop	001	010	1	1	Buffer Read, reset WIP				4T
trin,h trop,h trro,h	001	010	0	1	Buffer Read, reset WIP				2T
irop,b frop	001	001	1		Read Cycle	0	1	0	4T
trop,b trow,b	001	001	0		Read Cycle	0	1	0	3T
wrop,w wrro,w	010	011	0		Write Cycle - if BTACK - Write Cycle 2	0 0	1 0	1 1	4T 4T
wrop,h wrro,h	010	010	0		Write Cycle - if BTACK - Write Cycle 2	0 0	1 0	1 1	4T 4T
wrop,b	010	001	0		Write Cycle	0	1	0	4T

Figure 42.   Bus Access with AO(30:31) = 10 and SUPPRESS = 1.

and WIP is set, then the Bus Controller must have saved the previous halfword in its buffer. It runs a Write Cycle to transfer a full word, taking the upper halfword from its buffer and the lower halfword from the Processor. If the device does not accept the full word, the Bus Controller runs more Write Cycles to transfer the remaining data.

50

Figure 42 shows what happens in similar cases but with SUPPRESS active. Because Buffer Writes are inhibited, the WIP is never set for write requests.

5.2.4 AO(31) = 1. When the lowest bit of the byte address, AO(31), is 1, the data are aligned on an odd byte boundary. Figure 43 and Figure 44 show what the Bus Controller does in this case. Because of the odd address, there is no case involving a full word transfer. For word or halfword requests, the Bus Controller expects to run 2 bus cycles: Cycle 1 transfers the upper byte, using LDS (it is the lower byte of the halfword accessed by A31-A01); Cycle 2 transfers the lower byte, using UDS with A31-A01 = AO(0:30)+1. However, if the Bus Controller is reading with AO(30) = 0, and the device re-

When the Flow-charts say:	Then the PCR says:			Then the Bus Controller does:	And the Outside World sees:			And it takes at least:
Access	Cycle	Width	I	ICU Sequence	WS	UDS	LDS	
irin,w frin	001	011	1	Read Cycle, Error = Odd Instr Address	0	0	1	4T 3T
irop,w frop	001	011	0	Read Cycle if not WTACK - Read Cycle 2	0 0	0 1	1 0	4T 3T
trin,w	001	011	0	Read Cycle, Error = Odd Instr Address	0	0	1	3T 3T
trop,w trro,w	001	011	0	Read Cycle if not WTACK - Read Cycle 2	0 0	0 1	1 0	3T 3T
irin,h frin	001	010	1	Read Cycle, Error = Odd Instr Address	0	0	1	4T 3T
irop,h frop	001	010	1	Read Cycle if not WTACK - Read Cycle 2	0 0	0 1	1 0	4T 3T
trin,h	001	010	0	Read Cycle, Error = Odd Instr Address	0	0	1	3T 3T
trop,h trow,h trro,h	001	010	0	Read Cycle if not WTACK - Read Cycle 2	0 0	0 1	1 0	3T 3T
irop,b frop	001	001	1	Read Cycle	0	0	1	4T
trop,b trow,b	001	001	0	Read Cycle	0	0	1	3T
wrop,w wrro,w	010	011	0	Write Cycle 1 Write Cycle 2	0 0	0 1	1 0	4T 4T
wrop,h wrro,h	010	010	0	Write Cycle 1 Write Cycle 2	0 0	0 1	1 0	4T 4T
wrop,b	010	001	0	Write Cycle	0	0	1	4T

Figure 43. Bus Access with AO(30:31) = 01.

When the Flowcharts say:	Then the PCR says:			Then the Bus Controller does:	And the Outside World sees:			And it takes at least:
Access	Cycle	Width	I	ICU Sequence	WS	UDS	LDS	
irin,w frin	001	011	1	Read Cycle, Error = Odd Instr Address	0	0	1	4T 3T
irop,w frop	001	011	0	Read Cycle 1 Read Cycle 2	0 0	0 1	1 0	4T 3T
trin,w	001	011	0	Read Cycle, Error = Odd Instr Address	0	0	1	3T 3T
trop,w trro,w	001	011	0	Read Cycle 1 Read Cycle 2	0 0	0 1	1 0	3T 3T
irin,h frin	001	010	1	Read Cycle, Error = Odd Instr Address	0	0	1	4T 3T
irop,h frop	001	010	1	Read Cycle 1 Read Cycle 2	0 0	0 1	1 0	4T 3T
trin,h	001	010	0	Read Cycle, Error = Odd Instr Address	0	0	1	3T 3T
trop,h trow,h trro,h	001	010	0	Read Cycle 1 Read Cycle 2	0 0	0 1	1 0	3T 3T
irop,b frop	001	001	1	Read Cycle	0	0	1	4T
trop,b trow,b	001	001	0	Read Cycle	0	0	1	3T
wrop,w wrro,w	010	011	0	Write Cycle 1 Write Cycle 2	0 0	0 1	1 0	4T 4T
wrop,h wrro,h	010	010	0	Write Cycle 1 Write Cycle 2	0 0	0 1	1 0	4T 4T
wrop,b	010	001	0	Write Cycle	0	0	1	4T

Figure 44. Bus Access with AO(30:31) = 11.

sponds with WTACK, then it has delivered the entire halfword on D23-D08, and Cycle 2 is not needed.

There is one extra operation when the Processor tries to read an instruction that is not aligned on a halfword boundary. The Bus Controller reads the first byte, but instead of reading the second byte it reports the Odd Instruction Address exception.

5.2.5 Service Cycles. Service Cycle requests are similar to full word Bus Access requests. The Processor transfers first the upper and then the lower halfword in successive requests. The Bus Controller keeps track of the requests with its Service Word in Progress (SWIP) flag (see Figure 45). The data strobes provide additional

When the Flow-charts say:	Then the PCR says:		If SWIP is:	Then the Bus Controller does:	And the Outside World sees:			And it takes at least:
Access	Cycle	StSel		ICU Sequence	WS	UDS	LDS	
srak	101	101	0	Read Cycle, set SWIP	1	0	1	3T
srak	101		1	Buffer Read, reset SWIP				2T
srmc	101	bsb	0	Read Cycle, set SWIP	bsb			3T
srmc	101		1	Buffer Read, reset SWIP				2T
srpg	101	bsb	0	Read Cycle, set SWIP	bsb			3T
srpg	101		1	Buffer Read, reset SWIP				2T
swak	110		0	Buffer Write, set SWIP				3T
swak	110	101	1	Write Cycle, reset SWIP	1	0	1	4T
swmc	110		0	Buffer Write, set SWIP				3T
swmc	110	bsb	1	Write Cycle, reset SWIP	bsb			4T
swpg	110		0	Buffer Write, set SWIP				3T
swpg	110	bsb	1	Write Cycle, reset SWIP	bsb			4T

Figure 45. Service Cycles.

access information for Micro/370 Service Cycles. WS, UDS, and LDS form a 3-bit pattern which can be used to select various devices. The strobes can have values from 1 to 7, distinguishing Sense/1, Sense/2, (input) Service Cycles and Control/1, Control/2, ... Control/7 (output) Service Cycles.

Some Service Cycle requests have defined applications. Micro/370 uses Sense/3 and Control/3 Service Cycles to communicate with the coprocessors that execute decimal and floating-point instructions. It uses Sense/5 and Sense/4 to read the initial PSW required for IPL processing. It uses Sense/5 to read information needed to build an interruption code which must be stored for certain System/370 interruptions. Sense/7 and Control/7 fetch and store System/370 Storage Key data. Control/5 reports the new PSW (upper word via the Address Bus, lower word on the Data Bus) whenever PSWH is loaded (other than a simple change in the Condition Code). Control/6 signals a Dynamic Address Translation facility that the Purge TLB instruction is being executed.

DIAGNOSE instructions enable a control program to run any of the Service Cycles. The DIAGNOSE selects a Sense or Control Service Cycle and specifies the address to be placed on A31-A04 as well as the 3 strobe bits. The instruction specifies a register for the 32 bits of data transferred. Micro/370 runs the Service Cycle using SRPG or SWPG requests and returns the Bus Controller's completion status, which identifies the device's response.

```
┌─────────────────────────────────────────────────────────────┐
│  A Macrocycle is an uninterruptible sequence of Bus Accesses. │
│  It maintains the integrity of Read-Modify-Write and          │
│    PSW-Swap operations.                                        │
├─────────────────────────────────────────────────────────────┤
│                                                               │
│  The "Macrocycle In Progress" pin is activated ...            │
│                                                               │
│    • when the Flowcharts say:                                 │
│                    trop,b or trop,w or trow,b                 │
│      and the S/370 Instruction is:                            │
│                    CS, CDS, DIAG, LPSW, NI, OI, or TS          │
│                                                               │
│    • or when the Flowcharts say:                              │
│                    wrro                                        │
│                                                               │
├─────────────────────────────────────────────────────────────┤
│                                                               │
│  The "Macrocycle In Progress" pin is de-activated ...         │
│                                                               │
│    • when the Flowcharts say:                                 │
│                    a1                                          │
│                                                               │
├─────────────────────────────────────────────────────────────┤
│  Notes:                                                        │
│  1. 'a1': start executing a new S/370 instruction.            │
└─────────────────────────────────────────────────────────────┘
```

Figure 46. Macrocycles.

5.2.6 Macrocycles.   Certain System/370 instructions and PSW swaps (during interruption processing) require an uninterruptible sequence of bus cycles. They must read data, modify them, and then store the modified data without allowing any other device attached to the bus to access the same data. Micro/370 supports such read-modify-write operations with its Macrocycle.   Figure 46 explains how the Processor controls Macrocycles. During Macrocycles, the Bus Controller suspends bus arbitration, so that it does not grant mastery of the bus to any other device; it ignores solitary HALT input signals, which would ordinarily suspend Micro/370's bus cycles and allow other devices to use the bus; it rejects Bus Cycle Re-run requests unless they are accompanied by an overriding XRESET signal.

5.3 External Interrupts.

The Bus Controller synchronizes the 10 Interrupt Controls (see section "5.1.5 Interrupt Controls"). When an interrupt signal (one of the System/370 Interruption Requests or a specific decoding of the Interrupt Priority Level or Priority Request pins) has been detected during two successive clock cycles, the Bus Controller passes the signal via IRR to the Processor. State Machine Control responds to the signal as described in section "3.5 External Interrupts".

54

5.4 External Resets.

The Bus Controller synchronizes the 2 Reset Controls (see section "5.1.8 Reset Controls") and the HALT input. When an external reset signal (a specific decoding of the 3 signals RESET, XRESET, and HALT) has been detected during 10 successive clock cycles, the Bus Controller records it in XRR and asserts an internal reset signal, which kills any Bus Access operation that may have been in progress and tells the Processor's Clock-Phase Generator to not generate any phase outputs. After the external reset signal has been removed for 2 successive clock cycles, the Bus Controller removes its internal reset signal. Once the Clock-Phase Generator starts up again, State Machine Control responds to the signal in XRR as described in section "3.6 External Resets".

5.5 Bus Arbitration.

The Micro/370 Bus Controller is arbiter of the Micro/370 Bus. It decides when to grant mastery of the bus to an external device that requests it. The protocol is essentially the same as for Motorola M68000 architecture. The control pins are listed in section "5.1.6 Bus Arbitration Controls". The Bus Controller follows an algorithm for issuing Bus Grant similar to that of an MC68000 microprocessor. However, it adds to that algorithm some additional conditions which must be satisfied before it asserts Bus Grant. For example, it does not issue Bus Grant during Macrocycles, and it does not issue Bus Grant between the two Bus Cycles needed to transfer a halfword of data aligned on an odd byte boundary.

REFERENCES

1. IBM System/370 Principles of Operation, manual #GA22-7000-9 (IBM Corp., May, 1983).

2. Tredennick, Nick. How to Flowchart for Hardware. Computer, Vol. 14, No. 12 (December, 1981) 87-102.

MICROARCHITECTURE OF THE HP9000 SERIES 500 CPU

James G. Fiasconaro

Hewlett-Packard Co.
Systems Technology Operation
Fort Collins, Colorado 80525 USA

The HP9000 Series 500 computer uses a fully-integrated 32-bit processing system based on five custom NMOS VLSI circuits that operate at a frequency of 18 MHz. These chips include a 32-bit CPU, an I/O processor, a memory controller, a 128K-bit RAM, and a clock driver (1). The CPU and I/O processor are both microprogrammed. The I/O processor provides a high-performance input/output data path for the system peripherals. The I/O processor handles eight DMA channels and has an I/O bus bandwidth of 5.1M bytes/s when transfering at maximum rate. The memory controller chip can control 256K bytes of RAM and do single-bit error correction and double-bit error detection. A memory controller can map each of 16 16K-byte blocks of memory to its desired position in the address space and can "heal" up to 32 bad memory locations. The 16K x 8-bit RAM chip has redundant rows and columns for yield improvement and can handle pipelined memory accesses. The clock chip generates two nonoverlapping 18 MHz clock signals from a 36 MHz sine wave. A clock chip is used with each CPU, I/O processor, and memory controller/RAM subsystem.

CPU's, IOP's and memory controllers communicate via a system bus called the memory processor bus (MPB). The protocol of this 44-line, 36M-byte/s bus can support up to seven CPU's and IOP's and 15 memory controllers. This bus is multiplexed between 29-bit addresses and 32-bit data on alternating 55 ns clock cycles. Memory accesses are pipelined to allow sending up to two new addresses while the first data word is fetched from memory.

The remaining sections of this paper concentrate on the architecture of the CPU chip and on how the microcode controls the various parts of the chip. A brief description of the instruction set which the CPU chip executes is presented in order to illustrate how this instruction set affected the architecture of the chip. This paper does not address circuit design issues or other issues relating to the actual implementation, fabrication, or use of the chip.

1 OVERVIEW OF THE CPU CHIP

The major blocks making up the CPU chip (2) are shown in Figure 1. This figure both enumerates the blocks and shows the approximate chip area occupied by each of these major blocks. The sequencing hardware fetches words of microcode from the ROM which passes these words to the PLA. The PLA decodes the instructions (using a two-level NOR structure) and sends the appropriate control signals to all of the other blocks that make up the chip. The register stack is a set of general purpose and a few special purpose registers that are used by the microcode to implement the instruction set of the CPU. The ALU is the primary computational element of the chip. The MPB interface is the interface between the CPU and the system bus (Memory Processor Bus). The many qualifiers which the microcode needs to test in order to implement the CPU instruction set are generated by the various blocks that make up the CPU. The appropriate one is selected by the test condition multiplexer.

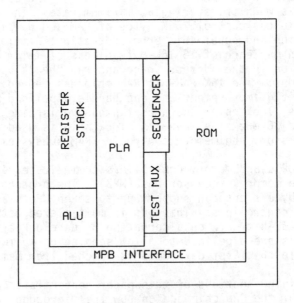

FIGURE 1.   MAJOR BLOCKS OF THE CPU

1.1 Overview of the Major Blocks of the CPU Chip

The ROM consists of 9216 38-bit words that are organized into 32-word pages. Each 38-bit word is referred to as a microinstruction. A new microinstruction can be fetched from the ROM every state (i.e. every 55 ns). The ROM is designed in such a way that skips and jumps within a 32-word page execute without interrupting the pipelined flow of microinstructions from the ROM to the PLA. A jump to another page is a two-state operation.

The sequencing hardware generates the 14-bit microinstruction addresses which go to the ROM. This piece of hardware provides the microcode with the ability to do short jumps (within a 32-word page), long jumps (to any ROM address), subroutine jumps (which can be nested three deep), subroutine returns, traps to subroutines, and skips. In addition, there is a mapper which converts the opcode of each CPU instruction into the ROM address corresponding to the beginning of the microcode for that CPU instruction.

Conditional jumps, skips, and traps in the microcode are executed based on the state of the so called test condition line which is the output of the test condition multiplexer. This multiplexer has 50 inputs which are generated by the various blocks making up the CPU. The input to be selected is specified by each microinstruction.

The register stack consists of 31 registers. There are two busses (called the A-Bus and the B-Bus) which run the length of the registers. Most of the registers can set from or dump to both of these busses. Some can set from and dump to only one of the busses. One of the busses (the A-Bus) connects the registers to the MPB interface. In addition, some of the registers are interconnected in such a way that they can transfer data to or recieve data from an adjacent register without using either of the main busses.

The ALU performs a viaiety of single-state, 32-bit arithmetic, logic, and shift operations. Operands are obtained either from the A-Bus and the B-Bus or from the ALU's internal busses. A shifter provides up to 31-bit right/left arithmetic or logical shifting. The logic functions AND, OR, and XOR are provided. The adder provides the sum, carry, and overflow for two's complement operands.

The MPB interface provides the mechanism by which the CPU communicates with memory and with other CPU's, I/O processors, and memory controllers. This interface contains seven 32-bit registers, and independent control logic. As a result, the CPU can initiate transfers and continue its own operation while the interface handles the MPB's synchronous protocol. The interface has dual-channel capability so that two different bus transactions can be in progress simultaneously.

The PLA implements two levels of logic in a NOR-NOR structure and has 55 inputs, 508 product terms, and 326 outputs. There is a small amount of extra hardware associated with the PLA which decodes the format of each microinstruction.

1.2 Microcode Control of the CPU Chip

The implementation of this CPU chip is based on a strategy of having a microcode routine in ROM for each instruction that the CPU must be able to execute. Furthermore, the microcode has complete responsibility for ensuring the proper utilization of all of the CPU hardware (except for those portions of the MPB interface which have separate control logic). Each microcode routine has three responsibilities over and above the proper implementation of a particular instruction. Each routine must keep the instruction pipeline full, test for external interrupts, and, at the proper time, execute the microinstruction which signals the sequencing hardware to go to the microcode routine for the next instruction.

The strategy for having the microcode control the hardware is as follows. The sequencing hardware fetches a microinstruction from the ROM and passes it to the PLA. The PLA decodes the instruction and sends the appropriate control signals to the rest of the circuits on the chip including the sequencing hardware. This interface between the PLA and the rest of the circuits is the key. It consists entirely of "control lines" going from the PLA to the rest of the chip and "qualifiers" going from the rest of the chip back to the PLA. Each piece of hardware is designed to do a specific operation when each of its control lines is asserted and to assert each of its qualifiers no later than a specified time after a control line is asserted.

Each microinstruction is 38 bits wide and contains two, five, or seven subfields as shown in Figure 2. The normal format is interpreted as follows. The TESTS field is a 6-bit wide field which, in general, selects one of the inputs to the test condition multiplexer. In most cases, if the test condition is true, then the next microinstruction is skipped. In some cases, if the test condition is true, the sequencing hardware will execute a trap (subroutine jump) to a location that is stored in a rom in the sequencing hardware. The SPECIAL field is a 6-bit wide field which either specifies an operation for some piece of the CPU or specifies that this microinstruction should be interpreted as one of the other formats. The A-BUS SOURCE field is a 6-bit wide field that specifies the register that should be dumped onto the A-Bus. The A-BUS DESTINATION field is a 6-bit wide field which specifies the register that should be set from the A-Bus. The ALU STORE

BITS 0-5	6-11	12-17	18-23	24-27	28-32	33-37	
TESTS	SPECIAL	A-BUS SOURCE	A-BUS DESTINATION	ALU STORE	ALU FUNCTION	B-BUS SOURCE	NORMAL
TESTS	LOAD OPERATION	A-BUS SOURCE	A-BUS DESTINATION	ALU STORE	ALU FUNCTION	CONSTANT	SHORT CONSTANT
B-BUS DESTINATION	LOAD OPERATION	A-BUS SOURCE	A-BUS DESTINATION	ALU STORE	ALU FUNCTION	CONSTANT	TRANSFER CONSTANT
B-BUS DESTINATION	TRANSFER OPERATION	A-BUS SOURCE	A-BUS DESTINATION	ALU STORE	ALU FUNCTION	B-BUS SOURCE	TRANSFER
TESTS	SHIFT OPERATION	A-BUS SOURCE	A-BUS DESTINATION	ALU STORE	SHIFT COUNT	B-BUS SOURCE	SHIFT
TESTS	JUMP	A-BUS SOURCE	A-BUS DESTINATION	ALU STORE	ALU FUNCTION	TARGET ADDRESS	SHORT JUMP
TESTS	LONG JUMP	A-BUS SOURCE	A-BUS DESTINATION	TARGET ADDRESS			LONG JUMP
B-BUS DESTINATION	32-BIT CONSTANT						LONG CONSTANT

FIGURE 2. FORMATS OF THE CPU MICROINSTRUCTIONS

field is a 4-bit wide field that selects one of the 16 possible combinations of the four registers inside the ALU to receive the ALU results. The ALU FUNCTION field is a 5-bit wide field which specifies the function that the ALU should perform. The B-BUS SOURCE field is a 5-bit wide field that specifies the register that should be dumped on the B-Bus. Note that in this format there is no B-BUS DESTINATION field.

Figure 2 shows four variations of the normal format that are signaled by the choice of the SPECIAL field. The first variation is the short constant format. In this case, the B-BUS SOURCE field is replaced by a 5-bit constant that is dumped onto the B-Bus. The second variation, the transfer constant format, is like the first variation except that in addition to that change the TESTS field is also replaced with a B-BUS DESTINATION field. The third variation, the transfer format, is like the normal format except that the TESTS field is replaced by a B-BUS DESTINATION field. The fourth variation, the shift format, is like the normal format except that the ALU FUNCTION field is replaced by a shift count for the shifter in the ALU. The SPECIAL field specifies left/right and arithmetic/logical for these shifts.

The short jump format is also shown in Figure 2. This format is signaled by the choice of the SPECIAL field. This case is similar to the normal case except that the B-BUS SOURCE field is replaced by a 5-BIT TARGET ADDRESS for the jump. The upper nine bits of the ROM address are not affected by this jump and so it can be used only for targets on the current 32-word page. If the test condition specified by the TESTS field is true then the jump will happen. If the test condition is not true then the next sequential microinstruction will be executed.

The next format shown in Figure 2 is the long jump format. This case is similar to the short jump format except that the ALU STORE, ALU FUNCTION, and 5-BIT TARGET ADDRESS fields are replaced by a 14-BIT TARGET ADDRESS field. This jump can be used to jump to any ROM address but it takes two states to execute this instruction instead of one state.

The final format shown in Figure 2 is the long constant format. This format contains only two fields, a 32-BIT CONSTANT which is dumped on the B-Bus and a B-BUS DESTINATION for this constant. This format can only be used immediately after a transfer format microinstruction.

2 THE CPU INSTRUCTION SET

In order to understand the detailed structure of the CPU it is necessary to understand some of the details of the instruction set

that the CPU executes. This instruction set has implications for
several parts of the CPU with the major impact on the registers.
The next two sections contain an overview of the CPU instruction
set and a discussion of the impact of this instruction set on the
CPU hardware.

2.1 Overview of the CPU Instruction Set

The CPU instruction set (3) is stack oriented. Each program
has its own execution stack for allocating local variables, passing
parameters to other procedures, saving the state of the CPU on
procedure calls, and evaluating expressions. There are instructions
for pushing data onto the stack from memory, and for popping data
from the stack and storing it in memory. Instructions get their
operands and parameters from the uppermost words in the stack and
leave their results on the stack.

A status register and an index register are also provided. The
status register contains information about the state of the program
that is currently running. Among other things, the status register
contains carry and overflow bits, a privilege level bit, an
interrupt enable bit, condition code bits, and the current code
segment number. The index register is used primarily in address
calculations and is treated as a 32-bit two's complement integer.
Both of these registers are saved in a stack marker (along with the
offset of the return point and the offset back to the next stack
marker) on every procedure call.

Segmentation is used to support virtual memory in the CPU
instruction set. Every program can use up to 4096 code segments
and 4096 data segments, and must use at least three segments -- a
code segment, a stack segment, and a global data segment. (Other
data segments used by a program are referred to as external data
segments and these may be paged.) Pointers to the smallest and
largest addresses occupied by each of these three segments are
refered to as program base, program limit, stack base, stack limit,
data base, and data limit as shown in Figure 3. In addition, the
program counter points to the current instruction in the (current)
code segment, and two other pointers are used with the stack
segment. Q points to the most recently pushed stack marker and S
points to the uppermost 32-bit word in the stack. The information
required to manage the segments used by each program is maintained
in memory-resident tables. Each program has its own code and data
segment tables and one common set of system code and data segment
tables is shared by all programs. There are no pointers to the
start and end of external data segments. Location and length
information for these segments is obtained from the appropriate
table in memory when it is needed by the microcode.

The instruction set provides a full repertoire of load and

62

store instructions for bit, byte, half-word, word (four bytes) and double-word quantities. All memory accesses using these instructions are bounds checked against program base and program limit, stack base and S, data base and data limit, or the location and length information in a table in memory. A bounds violation causes a trap to the operating system. Stores into code segments are not allowed. In unprivileged mode, a user can access only the user's own code, stack, global data, and external data segments. In privileged mode, any memory location can be accessed.

The primary integer data types supported by the instruction set are 16-bit and 32-bit two's complement integers, 32-bit

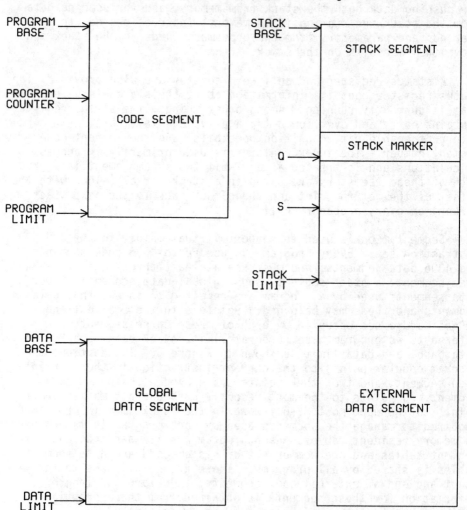

FIGURE 3.   SEGMENTS USED BY THE CPU INSTRUCTION SET

unsigned integers, and eight-digit unsigned decimal integers. The instruction set supports 32-bit and 64-bit IEEE-standard binary floating-point numbers and conversions between the 64-bit IEEE format to a 17-digit decimal floating point format. Both unstructured byte arrays and structured byte strings are supported. A wide variety of instructions are provided to manipulate all of these data types.

The primary instruction formats used in the instruction set are shown in Figure 4. Both 16-bit and 32-bit formats are used. Four types of address calculation are used for memory reference instructions. For direct addressing (I and X both zero), the effective address is simply base plus or minus offset. The 3-bit base field in the instruction specifies one of eight options: program counter plus offset, program counter minus offset, Q plus offset, Q minus offset, data base plus offset, data limit minus

16-BIT MEMORY REFERENCE

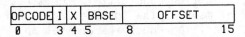

16-BIT CONDITIONAL BRANCH

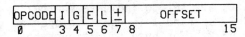

6-BIT OPERAND

STACK OPERATION / SPECIAL INSTRUCTION

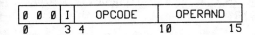

8-BIT OPERAND

32-BIT MEMORY REFERENCE AND UNCONDITIONAL BRANCH

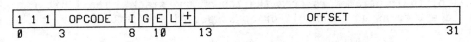

32-BIT CONDITIONAL BRANCH

FIGURE 4.   INSTRUCTION FORMATS IN THE CPU INSTRUCTION SET

offset, stack base plus offset, and S minus offset. For indirect addressing (I one, X zero), the location referenced by base and offset is read and interpreted as a pointer to the desired memory location. Indexed addressing (I zero, X one) is like direct addressing except that the 32-bit two's complement byte offset in the index register is added into the calculation. Indirect indexed addressing (I and X both one) is like indirect addressing except that the index register is added after the indirect pointer is evaluated in all cases except for a 32-bit unconditional branch instruction where the index register is added prior to fetching the indirect pointer. (This makes it easier to implement high level language CASE statements.)

The conditional branch instructions use the program counter as the base and provide only direct and indirect addressing. The bits labeled G, E, and L in Figure 4 stand for greater, equal, and less and any conbination of the three can be specified. The branch will be taken if the specified combination matches the 2-bit condition code in the status register. The condition code bits in the status register are set as a side effect of many of the instructions. For instructions like load or add the condition code bits are set to indicate whether the word loaded or the result of the add is greater than zero, equal to zero, or less than zero. For compare instructions the condition code bits are set to indicate whether operand 1 is greater than operand 2, operand 1 is equal to operand 2, or operand 1 is less than operand 2. For byte operations the condition code bits are set to indicate whether the byte is a special ASCII character, an upper or lower case alphabetic ASCII character, or a numeric ASCII character.

The 6-bit operand format is used primarily for shift instructions, byte array manipulation instructions, and conditional branch instructions that test other information such as the top-most word in the stack or the carry and overflow bits in the status register instead of the condition code bits in the status register. The I bit is used as an indirect bit, an index bit, or as a special indicator bit.

The 8-bit operand format is used primarily for arithmetic instructions that use an 8-bit immediate operand and the top-most word in the stack. The I bit is used as an indirect bit, an index bit, or as a special indicator bit.

The stack operation/special format is used for instructions which manipulate the top-most words in the stack or which obtain all of their parameters from the top of the stack. The I bit is used either as an index bit or as a special indicator bit.

2.2 Impact of the Instruction Set on the Hardware

Efficient implementation of the segmentation scheme used by
the instruction set requires that the pointers to the segments be
readily available to the microcode for use as base registers in
address calculations and for use as upper and lower limit registers
in bounds checking. This is accomplished by keeping these pointers
in dedicated registers on the CPU chip. There are separate
registers for program base, program counter, program limit, stack
base, Q, S, stack limit, data base, and data limit. These
registers are usually just general purpose registers but in a few
cases there is special hardware connected to the registers. The
program counter has an incrementer connected to it which can
increment by 2 or by 4 to accomodate 16-bit and 32-bit
instructions. In addition, there is a comparator between the
program counter and the program limit register to detect the case
of a program erroneously flowing past the end of the code segment.

Additional hardware support is provided for efficient address
calculation and bounds checking. The hardware decodes the base
field in the instruction so that the appropriate base register and
upper and lower limit registers can be accessed immediately by the
microcode independent of whether the current instruction is a
16-bit instruction or a 32-bit instruction. The hardware also
provides an easy way to access the 6-, 8-, or 19-bit offset from
the current instruction for the base plus or minus offset
calculation. The appropriate ALU operation, add or subtract, is
determined by the hardware. In some cases the offset in the
instruction is a halfword or word offset which can be shifted left
one or two bits to turn it into a byte offset. (All address
calculation is done with byte offsets.) Address calculation is also
made more efficient by hardware which provides easy tests of the I
and X bits in the current instruction. Two 1-bit bounds violation
flags are also provided along with a special add and a special
subtract operation for the ALU which can set or clear these flags.
These special ALU operations are used when performing bounds tests
on a computed address against the upper and lower limits. The
hardware also provides an easy way for the microcode to test if
either of the bounds violation flags is set.

The number of memory references required to maintain the
execution stack is greatly reduced by the addition of four more
special registers. These registers, referred to as TOSA, TOSB,
TOSC, and TOSD, are used by the microcode as a cache for the top
four words in the stack. TOSA would normally contain the top-most
word in the stack. There is also a "valid bit" for each of these
registers which indicates whether or not the register contains
valid data. These four registers are interconnected so that they
can be pushed and popped like a stack (the valid bits change

appropriately). The hardware provides tests of the validity of
each of these registers so that the microcode can maintain these
registers.

The use of this special hardware is best illustrated with
some examples. In a load instruction the microcode must read a
word from memory and push it onto the stack. The microcode does
this by first testing the validity of TOSD. For load instructions,
TOSD is usually not valid, so the microcode simply pushes this
stack of four registers and puts the word read from memory into
TOSA. If TOSD had been valid, the microcode would have written the
content of TOSD to memory, incremented the pointer to the
top-of-stack in memory, cleared the valid bit for TOSD, and then
proceeded with the load instruction as if TOSD had not been valid.
In a store instruction, the microcode must write the top-most word
in the stack to some memory location. The microcode does this by
first testing the validity of TOSA. For store instructions, TOSA
is usually valid, so the microcode simply writes TOSA to the
desired location and pops this stack of four registers. If TOSA
had not been valid, the microcode would have read the top-most word
in the stack from memory, decremented the pointer to the
top-of-stack in memory, and then written it to the desired
location. Note also that every time a word is read from or
written to the stack in memory tests are performed by the
microcode to detect stack underflow and stack overflow.

An instruction pipeline is provided by the hardware. This
pipeline holds instructions from the time they are fetched from
main memory until the instruction is completely executed. This
pipeline sends instruction opcodes over to the mapper in the
sequencing hardware which generates the ROM address corresponding
to the beginning of the microcode for that instruction. This
pipeline handles both 16-, and 32-bit instructions but it is
implemented in a way which forces 32-bit instructions to be
word-aligned in memory.

Two other general purpose registers have dedicated uses. One
is used for the index register and one is used to hold the pointer
to the current user's data segment table.

The status register is not a general purpose register. There
is special hardware to allow microcode to set and clear the carry
and overflow bits in this register and to set and test the
condition code bits as described earlier. In addition, the
microcode can test the "privilege bit" in this register to
determine the current mode of the machine. Some instructions can
be executed only in privileged mode.

3 DETAILED DESCRIPTION OF THE CPU

This section presents a detailed description of the architecture of the following major blocks of the CPU: the register stack, the ALU, the sequencing hardware, and the MPB interface. General descriptions of the ROM, the PLA and the test condition multiplexer have already been presented. More detailed descriptions of these pieces would be interesting from a circuit design point of view but not from an architectural point of view and so these pieces of the CPU are not discussed further.

3.1 The Register Stack

A detailed block diagram of the register stack is shown in Figure 5. This figure shows the registers, the A-Bus, the B-Bus, the opcode bus, and the microinstruction bus. The opcode bus connects the register stack to the sequencing hardware. The microinstruction bus connects the register stack to the ROM and PLA. An arrow pointing from a register to a bus indicates that the register can be dumped onto the bus. An arrow pointing from a bus to a register indicates that the register can be set from the bus.

There are eight "scratch pad" registers, labeled SP0 through SP7 in the figure. These registers are not visible to the CPU instruction set and are used by the microcode for such things as storing intermediate results, passing parameters to and recieving results from microcode subroutines, and so forth. SP5 serves as a buffer register to allow communication between the register stack and the sequencing hardware. SP5 is also a 32-bit shift register which can receive/send bit serial information from/to some special pads on the chip that are not used during normal operation of the chip but which were used during the development of the chip for debugging the hardware and microcode.

The instruction pipe is implemented with the three registers labeled CIR, NIR, and PIR. These abbreviations stand for current instruction register, next instruction register, and prefetched instruction register. Instructions can be shifted from PIR to NIR and from NIR to CIR without using the busses which go through the register stack. When an instruction is fetched from memory it goes into PIR. This pipe is shifted at the end of the microcode routine which implements each instruction. If NIR contains a 32-bit instruction, NIR is shifted into CIR and PIR is shifted into NIR. If NIR contains two 16-bit instructions, neither of which has been executed yet, then the left half of NIR is shifted into the right half of CIR and PIR is not shifted into NIR (which still contains one 16-bit instruction in its right half). If NIR contains contains one 16-bit instruction that has already been executed and one 16-bit instruction that has not been executed yet, then the right

68

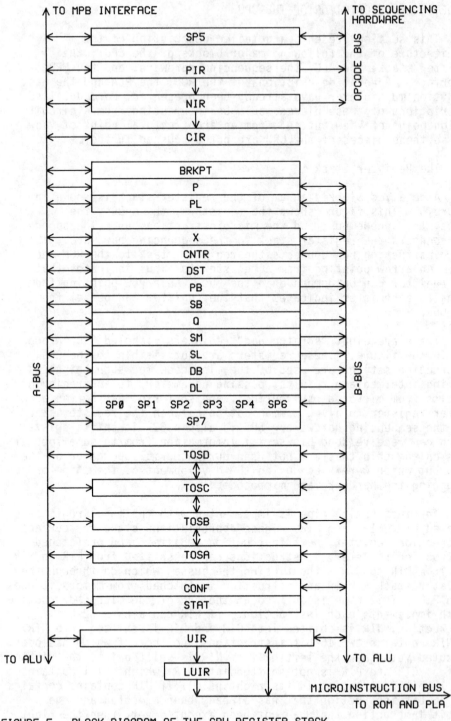

FIGURE 5. BLOCK DIAGRAM OF THE CPU REGISTER STACK

half of NIR is shifted into the right half of CIR and PIR is shifted into NIR. Thus 16-bit instructions are always in the right half of CIR when they are executed. This hardware does not handle the possible case where the left half of a 32-bit instruction is in the right half of a word in memory with the right half of the instruction in the left half of the next word in memory. This situation is avoided by having the language compilers and assemblers insert a NOP (no operation) instruction where necessary to force word alignment of 32-bit instructions. (The performance and code space penalties for this restriction are small because 16-bit versions of the most frequently used 32-bit instructions are provided.) The hardware for extracting the base field, the I bit, the X bit, and the GEL bits (for conditional branches) from the current instruction and the control lines for dumping the appropriate width offset (possibly shifted left one or two bits) onto the A-Bus are part of CIR.

The program counter (i.e. the address of the instruction that is in the current instruction register) is maintained in the register labeled P. The program limit pointer (i.e. the maximum legal address in the current code segment) is maintained in the register labeled PL. There is a comparator between P and PL which asserts a qualifier for the PLA if the two registers contain the same value. There is an incrementer circuit on the other side of P which can increment the value in P by two or by four. The output of this incrementer can be loaded back into P without using the A-Bus or B-Bus. There is one more register associated with the program counter. This is the breakpoint register (BRKPT). There is a comparator between the output of the incrementer attached to P and BRKPT which asserts a qualifier for the PLA if the register and the incrementer contain the same value. This is used for setting breakpoints in software.

The next set of registers in Figure 5 are primarily general purpose registers that have dedicated uses for this particular instruction set. The index register in kept in X. CNTR is a counter (i.e. a register with an incrementer attached to it) which can increment by one or by four. It has two primary uses. It is used to hold the address of the next instruction that is to be fetched from memory and loaded into PIR. It increments by four for this application. It is also used (usually in increment-by-one mode) as a loop counter by the microcode. In this case the instruction address is saved temporarily in a scratch pad register. A signal indicating the sign of CNTR is driven to the test condition multiplexer. The DST register contains the address of the current user's data segment table. The program base pointer (i.e. the minimum legal address in the current code segment) is kept in PB. The stack base pointer is kept in SB and the Q pointer is kept in Q. The register labeled SM contains the address of the top-of-stack in memory. The actual address of the top-of-stack as

seen by the CPU instruction set depends on how much of the stack is
in the registers TOSA through TOSD. The stack limit pointer is kept
in SL. The data base and data limit pointers are in DB and DL
respectively. The scratch pad registers SP0 to SP7 were described
earlier.

The next group of four registers are the top-of-stack
registers TOSA through TOSD. These registers can be pushed and
popped like a stack without using either the A-Bus or the B-Bus.
The use of these four registers is described in section 2.2.

CONF is the so called configuration register. This register
contains the valid bits for TOSA through TOSD, eight bits that are
used for high-level language degugging aids, and some one-bit
flags used by the microcode to indicate the existence of certain
situations.

STAT is the status register that is mentioned in sections 2.1
and 2.2.

The 38-bit microinstruction register (UIR) is used for
transfering microcode constants from the ROM to the register stack,
for reading the ROM, and degugging. The 12-bit register LUIR is
used to save the A-BUS SOURCE and A-BUS DESTINATION fields of each
microinstruction. There are cases where these fields must be
reexecuted as a result of an interaction with the MPB interface.
Like SP5, the microinstruction register is also a shift register
which can receive/send bit serial information from/to some special
pads on the chip that are not used during normal operation of the
chip but which were used during the development of the chip for
debugging the hardware and microcode. A register can be loaded
with a new value by shifting this new value into SP5, shifting the
appropriate microinstruction to transfer SP5 to the desired
destination into UIR, and then executing the microinstruction in
UIR. Similarly, a register can be read by shifting the appropriate
microinstruction to transfer the desired register to SP5 into UIR,
executing the microinstruction in UIR, and then shifting out the
value in SP5.

The register stack provides a variety of qualifiers for the
test condition multiplexer including: The most significant bit of
the A-Bus, the least significant bit of the A-Bus, an indicator
that the A-Bus is all zeros, the most significant bit of CNTR, an
indicator that the condition code in the status register matches
the GEL bits in the current instruction register, the I bit and the
X bit in the current instruction register, the valid bits for the
four top-of-stack registers, the privileged mode bit in the status
register, an indicator that P=PL, an indicator that the output of
the P incrementer is equal to BRKPT, two of the bits in the CONF
register, an interrupt pending signal, and an indicator that the

base field in the current instruction register is P-relative.

3.2 The ALU

The ALU performs a variety of 32-bit arithmetic, logical, and shift operations. A simplified block diagram of the ALU is shown in Figure 6. The ALU has two input registers (OPA and OPB in the figure), two functional blocks (the adder is one and the logical block and shifter make up the other one), and two output registers (OPC and OPD). The ALU gets its operands from and delivers its results to the register stack on the A-Bus and the B-Bus. There is also some hardware for doing BCD integer arithmetic, binary integer multiply operations, binary integer divide operations, address computation, and bounds testing.

The OPA and OPB registers receive data every state. The source of the data is specified by the microcode. OPA receives data either from the A-Bus or from the output of the adder. If the output of the adder is selected, the data is the result of the add

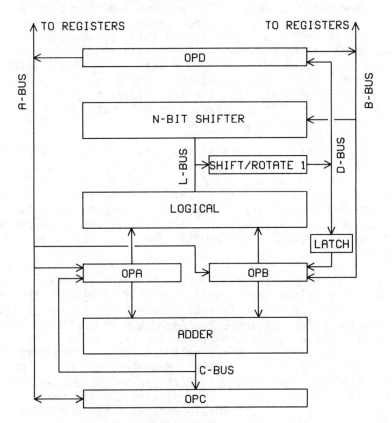

FIGURE 6. BLOCK DIAGRAM OF THE ALU

operation specified in the previous state. OPB receives data from the A-Bus, from the B-Bus or from the output of the logical/shifter block. If the output of the logical/shifter block is selected, the data is the result of the logical/shift operation specified in the previous state. The OPA register delivers both the true and complement of the data that is loaded into it. The OPB register delivers either the data that is loaded into it or zero. The choice depends on the function to be performed.

The logical block and shifter make up one of the functional blocks of the ALU. The logical block gets its inputs from OPA and OPB and either passes these inputs directly to the L-Bus or delivers a variety of AND, OR, and XOR functions to the L-Bus. This output goes to the SHIFT/ROTATE 1 block which either passes its input directly to the D-Bus or shifts/rotates the input by one bit left or right before putting the result on the D-Bus. A variety of 1-bit arithmetic/logical right/left shift operations and 1-bit right/left rotate operations are performed with this block.

The other major piece of this functional block is the N-BIT SHIFTER. This block consists of four cascaded shift left/right stages (shift 16, shift 8, shift 4, and shift 2) which drive the L-Bus and which can be combined with the SHIFT/ROTATE 1 block to produce any shift up to 31 bits in one state. The N-BIT SHIFTER gets its input directly from the B-Bus. The D-Bus can be loaded into OPD on the same state that the logical/shift operation is performed. The D-Bus can be loaded into OPB on the state after the logical/shift operation is performed.

The other functional block is the adder. This block performs a variety of 32-bit two's complement add and subtract operations (e.g. add OPA to OPB, subtract OPA from OPB, increment OPA, increment OPB, negate OPA, etc.). The adder receives its inputs from OPA and OPB and delivers the sum/difference to the C-Bus on the next state. The adder also provides two's complement overflow and carry information for the result. The C-Bus can be loaded into either OPC or OPA.

Because the adder can start a new operation on every state it is possible to pipeline adder operations as follows. On the first state the microcode can specify that a pair of registers should be dumped onto the A-Bus and the B-Bus (e.g. SP0 and SP1) and that an add operation should be performed. On the second state the microcode can specify that this result should be loaded into OPA and that a third register (e.g. SP2) should be loaded into OPB so that the sum of the first two registers can be subtracted from the third register. This process can be continued until the calculation is complete. Imtermediate results and the final result can be loaded into OPC, dumped on the A-Bus, and stored into the desired destination in the register stack.

Unlike the input registers, OPA and OPB, the output registers, OPC and OPD, are not loaded every state. Once loaded by the microcode, these registers hold their value until they are loaded again. OPC can set from and dump to the A-Bus. OPD can dump to either the A-Bus or the B-Bus but can only set from the D-Bus.

There are several other special functions performed by the ALU. The ALU saves interdigit carrys so that the adder can be used to add/subtract 8-digit wide BCD integers. The ALU is also set up to aid binary integer multiplication and division. There is a special multiply operation which performs the conditional add and shift for a multiply algorithm and a special divide operation which performs a selection and a shift for the implementation of a conditional subtraction divide algorithm. It is possible to compute a new bit of the product of two integers every state and it is possible to compute a new bit of the quotient of two integers every two states. The details of how this is actually done are beyond the scope of this paper. With proper attention to detail, these same functions can be used for floating-point multiplication and division as well.

The ALU aids address computation by providing an operation which is either an add or a subtract depending in the base field in the instruction in the current instruction register. The final special function of the ALU is bounds checking. The ALU has a special subtract instruction which sets a 1-bit bounds violation flag if the result does not carry and clears the flag if the result does carry. The computed address whose validity is being tested is loaded into OPA and the upper bound against which it is being tested is loaded into OPB. Similarly the ALU has a special add operation which adds the complement of OPA to OPB and sets another 1-bit bounds violation flag if the result carries and clears the flag if the result does not carry. The computed address is loaded into OPA and the lower bound is loaded into OPB.

The ALU provides a variety of qualifiers for the test condition multiplexer. Included in this are the following: The most significant bit of the C-Bus (i.e. the sign of the adder result), the ALU carry bit, the ALU overflow bit, and the NOR of the two bounds violation flags. (Note that these carry and overflow bits are different from the carry and overflow bits in the status register.)

3.3 Sequencing Hardware

The sequencing hardware is responsible for fetching the appropriate microinstruction from the ROM and passing it to the PLA for execution. In order to do this, it must provide the proper starting address for each microcode routine and properly execute all of the jumps and skips in the microcode. In addition, it must occasionally insert NOP (no operation) states because some of the microinstructions require more than one state for proper execution. A block diagram of the sequencing hardware is shown in Figure 7 which shows the major blocks which make up the sequencing hardware along with a block representing the ROM, the opcode bus which goes to the register stack, the microinstruction bus which goes to the PLA and the ROM address busses.

The opcode mapper recieves its inputs from NIR and PIR in the register stack. The choice depends on the mix of 16-bit and 32-bit instructions that the CPU happens to be executing. The opcode mapper is composed of two parts, an opcode decoder and a mapper rom. The opcode decoder converts the opcode field in a CPU instruction into an 8-bit address for the 256 x 14-bit mapper rom. The output of the mapper rom is the 14-bit starting address in the ROM of the microcode for a CPU instruction. The mapper is structured in such a way that both the opcode decoder and the mapper rom can be "reprogrammed" with a new set of CPU instruction set opcodes and a new set of start addresses with the same two mask levels that must be changed in order to reprogram the ROM. This flexibility permits changes to the CPU instruction set definition

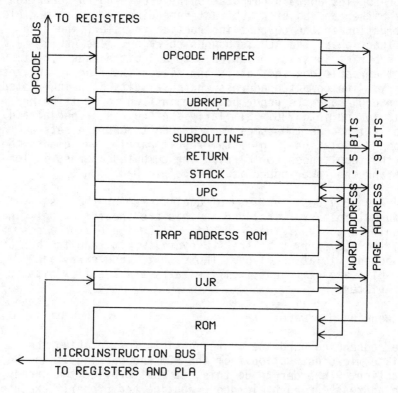

FIGURE 7. BLOCK DIAGRAM OF THE SEQUENCING HARDWARE

and permits the microcode to be moved around in the ROM as needed
for microcode updates. This was very helpful during the
development of the instruction set and microcode for the CPU chip.

The sequencing hardware contains a breakpoint register (UBRKPT
in the figure) for use in microcode debugging. If the breakpoint
register is enabled, then the CPU halts if the value in this
register equals the address on the ROM address bus. This feature
is not used during normal operation of the chip and this register
can be set only from the special pads on the chip that were
mentioned in connection with SP5 and the microinstruction register.

The microprogram counter (UPC in the figure) contains the
address of the next microinstruction to be fetched from the ROM.
There are two incrementers associated with this register, one is
used for normal sequencing and one is used for skips. There is
also a subroutine return stack associated with the microprogram
counter. When the microcode executes a subroutine jump, the
address of the return point is pushed onto this stack. When the
microcode executes a subroutine return, the return point is the
top-most address in this stack. This stack is three addresses deep
and it is up to the microcode to ensure that this limit is not
exceeded.

Normally the microcode must use the long jump format for a
subroutine jump instruction. However, a special case is made for
some of the most frequently used subroutine jumps (e.g. jumps to
the routines which manage the top-of-stack registers, jumps to the
routine which evaluates pointers, and jumps to the microcode part
of the interrupt handlers). These are called microcode traps. In
these cases, the normal format can be used and the TESTS field of
the microinstruction indicates that a trap should occur if the
condition is true. The target addresses for 13 traps are stored
in the trap address rom in the sequencing hardware. When a trap
occurs, the return address is pushed onto the subroutine return
stack as with any other subroutine jump.

The register labeled UJR in the figure is the microjump
register. The target address of a microcode jump is loaded into
this register from the microinstruction bus on the state before
the execution of the microcode jump. This register is invisible,
even to the microcode.

The sequencing hardware is also responsible for proper
sequencing of multi-state microinstructions. Examples of this
include: Jumps to arbitrary ROM addresses (2 states), a special
microinstruction which OR's the B-Bus and six bits of the A-Bus
into the next microinstruction (2 states), and a special
microinstruction which reads a specified ROM location and deposits
it into two scratch pad registers (3 states). A class of

microinstructions which can be either single-state or multi-state instructions includes those microinstructions which interact with the MPB interface. These are described in the next section.

3.4 The MPB Interface

A CPU communicates with the I/O processors, memory controllers, and other CPU's in a system via a system bus called the memory processor bus (MPB). The MPB interface is the link between this bus and the internal registers of a CPU. This interface is described in some detail in this section. This description includes a discussion of the operations that are possible on the MPB, an explanation of the architecture of the interface hardware, and a brief description of how microcode interacts with the interface hardware.

The operations that are possible on the MPB must be thought of in terms of address cycles and data cycles. Every CPU state (55 ns) is either an address cycle or a data cycle. These cycles alternate: address, data, address, data. This pattern of cycles is established when the system is powered up. Addresses and data share the same 32 physical lines on the bus with addresses being transfered on the bus during address cycles and data being transfered on the bus during data cycles. Every operation on the bus uses one address cycle and one data cycle. An operation that uses address cycle N will use data cycle N+5. The intervening address and data cycles may be used for other transactions.

The MPB is one area where the limited pad count available for VLSI chips becomes apparent. This bus is made up of very few physical lines. There are 32 address/data lines, 8 lines used by the processors on the bus to request access to the bus, and only four other control lines. The first of these control lines indicates whether or not a valid address is on the bus during address cycles and whether or not valid data is on the bus during data cycles. The second one indicates read or write during address cycles and double bit error during data cycles. The third one indicates whether each cycle is an address cycle or a data cycle. The final one is used to report error conditions back to the processor originating the bus transaction.

The reason why it was important to keep the number of physical lines on the system bus small is not apparent from looking at the CPU but is apparent from looking at either the I/O processor or the memory controller. These latter two chips must connect to two busses while the CPU must connect only to the MPB. The I/O processor is connected to a 32-bit I/O bus and the memory controller is connected to a 51-bit memory bus. The end result (after adding pads for power and ground, clocks, and testing) is that each of these two chips has in excess of 100 pads.

Eight operations are defined for the MPB. The upper three bits of the address lines specify the type of transaction while the remaining 29 bits specify an address. Four operations are memory transactions and four operations are inter-processor transactions. The four memory transactions are word (i.e. 32-bits), halfword, byte, and semaphore transactions. The direction of the transaction, read or write, is specified by one of the 12 control lines that are part of the MPB. Byte, halfword, and word reads and writes perform the obvious memory operation. Read data is right justified by the memory controller, write data is right justified by the CPU and positioned appropriately in the designated memory location (32 bits) by the memory controller. A semaphore transaction is an uninterruptable "word read and set to all ones".

The four inter-processor transactions are primary mode, secondary mode, directed message, and broadcast message transactions. The read/write control line has meaning for the primary and secondary mode transactions. Message transactions are effectively writes from one processor to another processor or from one processor to all processors. For each of these transactions, the upper three bits of the 29-bit address field contains the "MPB channel number" of the processor that should receive the transaction and the next four bits contain the memory controller number of the memory controller that should receive the transaction, if appropriate. (At power up, each CPU and I/O processor is assigned a unique 3-bit MPB channel number and all of the memory controllers are assigned to channel zero but each is assigned a unique 4-bit memory controller number.) Primary and secondary mode reads and writes are operations from and to the so called "slave address" and "slave data" registers in the MPB interface of another processor. The two modes, primary and secondary, are arbitrary but are intended to be used in a certain way; namely, a processor should stay in primary mode unless it wishes to establish exclusive communication with one other processor and prevent other processors from reading from or writing to its "slave" registers for the duration of time that it is in secondary mode. For message transactions, the bottom 25 bits of the address get OR'ed into the corresponding bits of the so called "message register" of the destination processor(s). Message transactions are used primarily for I/O interrupts and for system error indications.

A block diagram of the registers that make up the MPB interface is shown in Figure 8. These registers can be divided into three sets. The first set, the master registers, consists of the master address register, the X data register, and the Y data register. These registers are used by the CPU to initiate transactions on the MPB. For example, when the microcode reads from memory, it loads the desired address into the master address register, indicates the size (byte, halfword, or word) of the transaction, indicates which data register (X or Y) should get the data, and signals the interface to complete the transaction. The microcode is then free to do whatever it wants to do. The interface will gain access to the bus (based on a priority scheme that gives the processors with the lower MPB channel numbers higher priority but does not allow any processor to use all of the bus bandwidth), issue the address on the bus at the proper time, and put the data from memory into the designated data register when it arrives. When the CPU is ready to use the data it executes a microinstruction to dump the designated data register onto the A-Bus. If the data has already arrived from memory, then the dump occurs and this is a one-state microinstruction. If, however, the data has not yet arrived from memory, then the sequencing hardware "hangs" and repeats the A-BUS SOURCE and A-BUS DESTINATION fields of the microinstruction until the data arrives from memory. This is an example of a multi-state microinstruction whose execution time cannot be predicted.

Write operations are similar but there are some noteworthy differences. In this case, the microcode loads the master address register with the desired address, indicates which data register should be used, loads the data into this register, and signals the interface to complete the transaction. As with reads, the microcode is then free to do whatever it wants to do while the interface completes the transaction. Because the memory is word oriented with error detection and correction bits for the whole

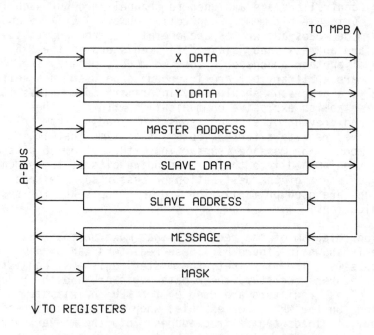

FIGURE 8. REGISTERS IN THE MPB INTERFACE

word (and not for each byte) write operations are implemented by reading the appropriate word, merging in the new bytes, computing the new error correction bits, and writing this new word into memory. When the CPU initiates a write, it is actually initiating the read half of the operation. The memory controller steals an address cycle/data cycle pair to do the write half of the operation. This scheme causes problems, however, if a write to some memory location is followed immediately (i.e. in the very next address cycle) by another memory transaction involving the same word in memory. This situation is avoided by the interface hardware which is designed to stay off of the bus immediately after initiating a write unless the microcode has indicated that it is alright to use the bus in this manner (i.e. the microcode knows that the second memory transaction is guaranteed to be to a different memory location).

This interface between the chip and the system bus relies heavily on so called "hang conditions" for proper operation. The interface provides qualifiers to the sequencing hardware to suspend execution of the microcode in the following situations:

1) The microcode is attempting to load the master address register when this register already contains an address which has not been issued on the bus yet.

2) The microcode is attempting to dump the X or Y data register when this register has not yet received data from the bus.

3) The microcode is attempting to load the X or Y data register when this register already contains data which has not been written on the bus yet.

This scheme permits the microcode to overlap many activities with memory transactions.

Because the system bus can start a new transaction on every address cycle, it is possible for the microcode to overlap memory accesses. For example, the microcode can start a read from memory and before asking for the data for this read start two other read operations. In the simplest case, the three addresses go onto the system bus on three consecutive address cycles and the three data words return from memory on three consecutive data cycles. This type of operation is very useful for reading short blocks of memory such as stack markers and the top-most words of the stack in memory. However, considerable care is required on the part of the microcode because there is no mechanism in the hardware to prevent data in one of the data registers (e.g. the data for the second of three reads) from being written over (e.g. by the data for the third of three reads) before it is dumped by the microcode. There are rules which the microcode must follow to ensure that this does not happen.

The second set of registers in the MPB interface (see Figure 8) are the slave registers. These registers are used for transactions that are started by some other processor (which uses the master registers of its MPB interface). The slave address register receives the address half of the transaction. For writes, the slave data register receives the data from the other processor. For reads, the slave data register supplies the data to the other processor. These registers are used by the CPU to communicate with individual I/O processors. For I/O writes and I/O commands the CPU writes to the slave registers of the I/O processor. If the slave registers of the I/O processor already contain valid data, then the data from the CPU is thrown away and the MPB interface reports this as an error condition to the CPU which can then retry the transaction. If the CPU wants to read from the I/O processor then the CPU puts itself into secondary mode and writes a command to the I/O processor telling it what information is needed. The I/O processor writes the data back to the slave registers of the CPU using secondary mode so that the data will not be rejected by the MPB interface on the CPU.

The final set of registers in the MPB interface are the message registers. The right-most 25 bits of the message register are set by message transactions on the system bus. The right-most 16 of these 25 bits are used for 16 I/O interrupt levels. The remaining of these 25 bits are used for reporting various system error conditions. A message transaction can be used by a processor to OR a one into any (or all) of these 25 bit positions. The CPU microcode can clear but not set bits in the message register. The remaining bits in this register have dedicated uses. A one is OR'ed into the most significant bit if the memory system detects a double bit error. The next two bits are used by the MPB interface to report error conditions such as an attempt to access a nonexistent memory location, or an attempt to write to another processor's slave registers while these registers already contain valid data. The next three bits of the message register are always zero. A one is OR'ed into the next bit when another processor writes into the slave data register on the CPU.

The final register is the mask register. This register is used to temporarily disable bits in the message register. The message and mask registers cause interrupts for the CPU in the following manner. If a bit in the message register is a one and the corresponding bit in the mask register is a zero then an interrupt condition exists. (Note that there is an interrupt enable bit in the status register which must also be a one before the CPU microcode will acknowledge an interrupt.) The bits in the mask register corresponding to the bits in the massage that are always zero are used to hold the MPB channel number assignment for the CPU.

The MPB interface provides two qualifiers to the test
condition multiplexer. The first is an indication that some
processor has written data into the slave data register. The
second is an indication that the master section of the MPB
interface is "active". This qualifier is true from the time the
microcode initiates a transaction on the bus until the transaction
actually is completed (which may be much later if the bus is very
busy).

4 SUMMARY AND ACKNOWLEDGEMENTS

This paper has described the microarchitecture of the CPU for
the HP9000 computer (i.e. the architecture of the chip from the
point of view of someone writing microcode for that chip). This
description is structured around the major blocks that make up the
CPU: the register stack, the ALU, the sequencing hardware, the
system bus interface, the ROM, the PLA, and the test condition
multiplexer. The machine instruction set of the CPU (i.e. the
architecture of the chip from the point of view of someone writing
assembly language software for the HP9000) has been described also
in order to show how this instruction set affected the
microarchitecture of the chip.

The CPU chip and the other VLSI chips that were developed for
the HP9000 are the result of the work of many dedicated employees
of the Hewlett-Packard Company. It would not be practical to
acknowledge the contribution of each one at this time. A summary
of the contributions of this group of people to the development of
this chip set is available in the August, 1983 issue of the
Hewlett-Packard Journal. The articles in this issue of the Journal
provide more detail on all of the chips, present information on the
IC process that is used to manufacture the chips, and describe the
packaging technology used for this chip set.

5 REFERENCES

1. Beyers, J.B. et al. VLSI Technology Packs 32-Bit Computer
   System into a Small Package, Hewlett-Packard Journal 8 (1983)
   3-6.

2. Burkhart, K.P. et al. An 18-MHz, 32-Bit VLSI Microprocessor,
   Hewlett-Packard Journal 8 (1983) 7-11.

3. Fiasconaro, J.G. Instruction Set for a Single-Chip 32-Bit
   Processor, Hewlett-Packard Journal 8 (1983) 9-10.

THE DRAGON COMPUTER SYSTEM
An Early Overview

Edward M. McCreight

Xerox Corporation
Palo Alto Research Center
3333 Coyote Hill Road
Palo Alto, California 94304

1 INTRODUCTION

Dragon is a new computer system being designed at the Xerox Palo Alto Research Center. In many ways Dragon is the newest descendant in the line of research personal computers that began in 1973 with the Alto and ends today in the Dorado. Like those machines, Dragon will be a personal computer. To us, this means a high-performance display controller and a virtual memory with only rudimentary protection hardware. Dragon and Dorado both execute programs compiled from the Cedar language. (Cedar is an extension of the Xerox Mesa language, which is a derivative of Pascal.) We plan to build modest numbers of Dragons for ourselves to allow us to explore useful ways of exploiting such performance in a personal setting before such machines are economical in widespread use.

Dragon differs in significant ways from its predecessors. Most importantly, it is a tightly-coupled multiprocessor. We see multiprocessing as an important wave of the future, and we need to understand its implications for software design. Dragon is also a 32-bit machine, both in its arithmetic path and in its virtual addresses. It has a new instruction set, considerably reduced in semantic complexity from previous machines. Dragon can handle 16-bit data, but at a noticeable penalty in code size and execution speed compared with 32-bit data. These facts will encourage our systems programmers to eliminate vestiges of 16-bit-ness that survive from Alto days. Dragon is also our first machine to include custom integrated circuits; previous machines were built of commercial integrated circuits.

Dragon is not even close to being finished. Our logic simulations are about two-thirds complete, and floor planning and layout have begun on some chips, but modest architectural

84

refinements are still a weekly event. We can fairly claim to have made a solid beginning, but a mountain of hard work remains. We hope by publishing this paper to stimulate discussions that could make Dragon a better machine.

2 KEY IDEAS OF DRAGON

Our present workhorse machine, the Dorado, is a personal computer in the same sense that a Rolls-Royce is a personal automobile. It is expensive, power-hungry, and noisy. Those facts alone would be enough to make us want a better machine. But we also had some new ideas that we wanted to investigate through implementation.

2.1 Multiprocessing

The first idea is tightly-coupled MIMD (multiple instruction streams operating on multiple data streams) multiprocessing. As we considered the declining fraction of total system cost represented by a single processor, the slow advance in processor circuit speed, and the knee in every processor speed vs. cost curve, it became apparent to us that the most straightforward way to obtain an order-of-magnitude increase in processing power was not to attempt an order-of-magnitude increase in circuit speed or architectural complexity, but rather simply to use multipleprocessors.

This approach has two major drawbacks. The first is that as the number of processors increase, so does the difficulty of routing information among them in general ways. If the processors are all attached to a common shared memory bus, then access to that bus becomes the resource that limits system performance. A partial solution to this problem is to interpose a cache memory between every processor and the shared memory bus. Until recently this solution has suffered from the "cache consistency problem." When a processor stores new data into its cache, should that store always appear on the shared memory bus? If so, serious contention results, because typically one-third of memory operations are stores. If not, then what prevents two caches from having different opinions about the contents of the shared memory? Fortunately a new architectural element, the two-port cache, reduces each processor's average need to store to the shared memory bus, thereby allowing modest numbers of processors to coexist without crippling interference.

The second drawback is more serious, especially in a personal computer. To take advantage of Dragon-style multiprocessing, there must usually be several processes ready to run, and the interprocess communication and synchronization among processes must be modest. While this is the normal case for a heavily-loaded timesharing system, maintaining many ready processes in a personal computer requires breaking up such compute-bound programs as compilers (and many others) into processes. Our existing compilers, for example, were written as single processes because we knew that they would be executed on single-processor systems. To have written them otherwise would have complicated the programming task and made the compiler run more slowly.

Dragon introduces us to two interesting software problems. First, how should we adapt existing software for the new multiprocessing hardware? Second, how should the coming wave of multiprocessor hardware change how we write programs, or how we think about programming? We at Xerox suppose that it is harder to write programs as multiple processes than as single processes, but that supposition is based more on second-hand reports than on first-hand experience. We intend to use Dragon as a tool for investigating how Dragon-style multiprocessing can be used to best advantage.

2.2 Faster Execution

Speed isn't everything, but it's kilometers ahead of whatever is in second place. Processor speed can be put to use directly in those few places in a program where it's needed, and in other places traded away for clarity of expression, compactness of representation, an extra level of interpretation, delayed binding, garbage collection, or a host of other worthwhile things. We hoped to improve the speed of Dragon, compared with our earlier machines, primarily in two areas. The first is serial execution of basic blocks, where pipelining and rich pipeline bypass paths reduce the average number of cycles per instruction. The goal, like that of the IBM 801, was a simple instruction set and a storage hierarchy that would permit the machine to execute an instruction nearly every machine cycle. The second area of improvement is the procedure call, where simplified parameter passing, a cached control stack, and elimination of several indirections in common cases would provide a dramatic speed improvement over our earlier implementations.

2.3 Code Density

One of the most notable achievements of the Mesa machine instruction set, implemented in microcode on such machines as the Dorado and the Xerox Star, was the compactness of compiled Mesa programs. As one extreme example, Baskett's Puzzle benchmark compiled into only one-fourth as many code bytes in an early version of the Mesa instruction set as in the IBM 370/4331 instruction set. This compactness has two benefits. The most obvious benefit is that a given program could be stored in fewer dollars' worth of memory. That benefit will remain significant for some time to come, since the declining cost per byte of memory has been nearly balanced by the need to store more ambitious programs. A less obvious benefit is that the smaller a program is, the less bandwidth is required to carry it from one place to another: from disk to memory, from memory to cache, from cache to instruction decoder.

Our challenge was to make the Dragon go faster, and to deal with a much larger address space, without throwing away the compactness that the Mesa architects had worked so hard to achieve.

3 THE M-BUS

Figure 1 shows an overview of a typical Dragon system. The central data pathway is the M-bus. We may eventually design Dragon systems with more than one M-bus, but for now there is one M-bus in a system. The M-bus is a synchronous time-multiplexed bus that carries data among the processor caches, the main memory, the map processor, and the display controller. M-bus mastership is assigned by a single round-robin arbiter. M-bus transactions are time-sequenced on a 32-bit address/data bus as dictated by the operation specified by the current M-bus master. Addresses on the M-bus are real (not virtual), and there are separate 32-bit address spaces for I/O and memory.

86

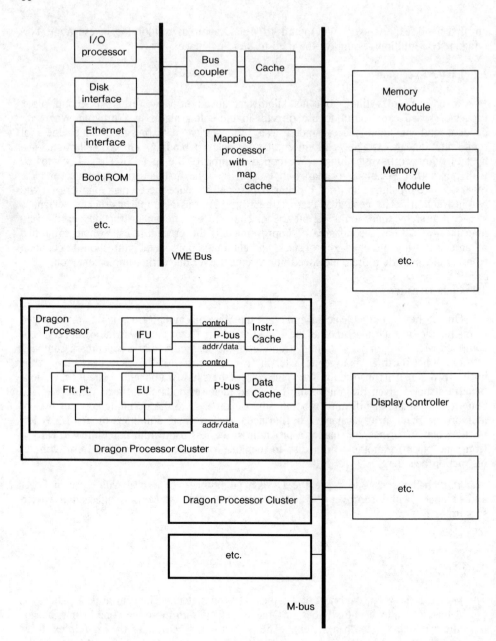

Dragon System
Figure 1

The following M-bus transactions are defined. A single line in the description below takes one machine cycle. Master commands are in **boldface**. Slave responses are in lightface. Address/data bus contents are in *italics*. Flow control is the insertion of optional wait states in the protocol by a slave that needs more time; w refers to the number of wait states inserted.

i. ReadQuad reads four words from memory or a willing cache. This cyclic order transports the requested data word first, so the requesting processor can continue after cycle w + 1.

    0: **ReadQuad** *address*

    {anything but ReadQuadReady}*

    w + 1: ReadQuadReady *data[address]*

    w + 2: *data[address-(address MOD 4)+(address+1 MOD 4)]*

    w + 3: *data[address-(address MOD 4)+(address+2 MOD 4)]*

    w + 4: *data[address-(address MOD 4)+(address+3 MOD 4)]*

ii. WriteQuad command writes four words to memory. The address must be equivalent to zero MOD 4. If the memory asserts WriteQuadReady in cycle 4 as shown, then the next transaction can commence in the next cycle. Otherwise the next transaction waits until the cycle after the memory asserts WriteQuadReady. Flow-controlling after data transport requires the memory to have a four-word write data buffer, but it allows more time for the memory to decide whether wait states are needed.

    0: **WriteQuad** *address*

    1: *data[address]*

    2: *data[address+1]*

    3: *data[address+2]*

    4: WriteQuadReady *data[address+3]*

iii. WriteSingle broadcasts a single word in the memory address space. The Master cache does this because it wants to change a word, and the its Shared flag (to be described shortly) for that word is set, indicating that a copy of that word may exist in other caches. This word is **not** written to memory, and there is no flow control.

    0: **WriteSingle** *address*

    1: *data[address]*

iv. IORead reads a single word in the I/O address space.

    0: **IORead** *address*

    {anything but IOReadDone}*

    w + 1: IOReadDone *data*

v. IOWrite writes a single word in the I/O address space. If the addressed Slave responds with IOWriteDone in cycle 1 as shown, then the next transaction can commence in the next cycle. Otherwise the next transaction waits until the cycle after the Slave responds.

    0: **IOWrite** *address*

    1: IOWriteDone *data*

Interprocessor atomicity is implemented by allowing the M-bus master to lock the arbiter. There is no automatic timeout, so processors (or caches) must be designed so they cannot lock too long.

3.1 Caches

Dragon caches, like other caches, have two related purposes. A cache provides its processor with faster access to most frequently accessed data than the memory system could. This is doubly effective in Dragon's case, because even when a Dragon cache does not contain the data requested by its processor, it can often implement by itself the virtual-to-real mapping that is required in order to find that data. It can do this because our solution to the cache consistency problem requires that a Dragon cache entry contain not only virtual address and data, but real page number as well, which is nearly all the information in the mapping table entry. By adding the rest of the mapping information to each cache entry, we enabled our cache to map any virtual address for which it contains an entry in the same page.

The second purpose of the caches is to reduce the M-bus bandwidth needed by a single processor. This is the only way that modest numbers of processors can share a common M-bus without crippling mutual interference.

3.1.1 Cache Architecture. A Dragon cache intermediates between a processor's P-bus (the cache's "front" door) and the system's common M-bus (the cache's "back" door). A cache contains 64 associative entries, each of which comprises the virtual page number, the real page number, the address of the block within the page, the block of 16 32-bit words of data, and various property bits for the page and for each group of four words within the block.

A cache's P-bus consists of a 32-bit phase-multiplexed virtual address/data field, a parity bit, a command field, a ready bit, and an error field. There is only one processor on a P-bus, and it is always master. One or more caches, and the floating-point accelerator chip(s), are slaves on the P-bus. A P-bus slave decodes its instructions from a combination of the P-bus command and virtual address fields. Commands to caches include { Normal | Locked | IO }{ Read | Write }. A slave responds by setting the ready bit and the error field. When a cache contains data being read or written, it asserts ready in the phase following the read or write operation, so that in the best (and, we hope, most likely) case, a cache read or write takes one machine cycle. Cache errors include PageFault and WriteProtectFault.

At the moment, the cache replacement algorithm replaces entries in strictly sequential order without regard to whether their data is dirty, although this may change. A cache's command decoding can be restricted to a subset of virtual addresses via a bit-serial initial setup path. This mechanism allows multiple cache chips to be attached to a P-bus, thereby implementing a larger but set-associative cache, and improving the hit ratio.

3.1.2 Cache Consistency. A system of caches is consistent, or coherent, if all copies of the value in each memory location are identical. An obvious way to maintain consistency is to broadcast all stores to the M-bus. Since typically one-third of data references are stores, at best this solution reduces the M-bus bandwidth requirements of a processor by a factor of three, independent of cache size. This factor is not large enough for Dragon.

A better solution is to broadcast on the M-bus only those stores that affect data existing in more than one cache. Dashiell and Thacker at Xerox were among the first to discover a way to implement this solution, and Goodman at the University of Wisconsin discovered a similar one at about the same time (2). Both solutions depend on having a second associator in each cache, called the back-door associator, or the "snoopy" associator. This second associator monitors M-bus reads, and volunteers data that it knows. If the data comes from another cache, instead of from the main memory, then all caches involved mark their copies as shared. Any change to cache data that is marked shared is broadcast to the M-bus.

The Dragon implementation uses a SharedData line in the M-bus, and a Shared flag in each cache entry. During the ReadQuad M-bus transaction, if the real address of some entry in a non-master cache matches the real address on the M-bus, that cache sets that entry's Shared bit, asserts SharedData on the M-Bus, and provides that entry's data during data transport. During WriteSingle (caused by a processor storing to a cache entry whose Shared flag is set), if the real address of some entry in a slave cache matches the real address on the M-bus, that cache asserts SharedData, and overwrites that entry's data during data transport. The master cache entry always loads its Shared flag from the value of the SharedData M-Bus line. In this manner a cache is constantly aware of shared data, and eventually discovers when sharing stops.

3.1.3 Cache Performance. The ultimate goal is for the entire system to execute a set of programs as fast as possible. System performance is determined by the interaction of processors, caches, and M-bus, so we cannot discuss cache performance in isolation. Although we haven't talked much about a Dragon processor yet, we concentrate our attention on how our caches behave when attached to a Dragon processor, since we believe that that behavior will have the greatest influence on total system performance. Other contributions will come from other processors: the map processor, the display controller, and the I/O processor.

A Dragon processor attaches to two caches. It fetches its instructions from one, and fetches and stores its data in the other. We expect these two caches to differ significantly in such performance statistics as hit ratio and average machine cycles between references, because the processor uses the two caches quite differently. At the moment each Dragon cache is planned to contain 64 entries, each entry including a virtual address (divisible by four words), a comparator, and four data words. That is a fairly small cache, and we are considering enlarging it, but for definiteness we will assume that size throughout the rest of this section.

Our own data about cache performance are from two sources. One is a cycle-by-cycle Dragon simulation. From this simulation we know quite a lot about the Dragon's performance on a few simple benchmark programs, notably Baskett's widely-circulated Puzzle. Some statistics for Puzzle are likely to apply as well for more "typical" programs (like compilers or text formatters), while others probably will not. We expect that Puzzle's average of 7 machine cycles per data cache reference and 1.8 machine cycles per instruction cache reference is fairly robust. On the other hand, we expect that cache hit ratios will vary considerably from benchmark to benchmark. As an extreme example, the Puzzle instruction cache hit ratio was very high because nearly the entire program fit within the instruction cache. One would not expect such behavior from a more complex program.

Ideally, cycle-by-cycle simulation would tell us every performance statistic we need. Practically, it suffers from two limitations. Our Dragon Cedar compiler is not ready, so not much Dragon code exists, and what does exist is hand-coded. Furthermore, even if we had plenty of interesting programs expressed as Dragon code, our simulator's speed limits the number of instructions we can simulate.

Our other performance data are measurements taken by N. Suzuki two years ago. He modified our standard Dorado microcode to collect statistics on Dragon cache performance. He pretended that the Dragon was executing the Dorado Mesa instruction set, and he collected statistics only for data references.

The two kinds of performance data make quite different assumptions, so the fact that they agree very closely on Puzzle's average data cache hit ratio (89.7% vs. 89.8%) should be regarded as accidental. Suzuki's data also indicates that the Mesa compiler (a "typical" program) has a data cache hit ratio that is 7% worse than Puzzle. This suggests that a "typical" program might achieve an average 83% data cache hit ratio. If we assume that a cache operation happens once every seven cycles, and that the average cache miss costs six cycles, then Dragon should run 14% slower on a "typical" program with our data cache than with an infinite data cache that never misses.

Performance will likely be worse for the instruction cache than for the data cache. Unfortunately, at the moment all our instruction cache performance data for "typical" programs is second-hand. J. Smith and J. Goodman (11) measured average instruction cache hit ratios over three programs, NROFF, CACHE, and COMPACT, all compiled from C and running on a VAX-11 under UNIX. Their results indicate a 90% hit ratio for a Dragon cache. Based on their measurements, they argue that cache size is the most important factor in determining hit ratio, and that for caches the size of ours, random replacement is better than LRU or FIFO, and direct mapping is nearly as good as full associative lookup. D. Patterson, et al., measured the hit ratio of a simulated RISC II executing the Portable C compiler with a cache similar to (but with a narrower entry than) ours (7). Their results indicated a hit ratio of 87%. This suggests that our instruction cache hit ratio will be somewhat better than our data cache hit ratio. On the other hand, we will likely spend many more cycles waiting for instruction cache misses than for data cache misses, because our simulation indicates that Dragon accesses the instruction cache almost four times as often as the data cache.

If our cache hit ratios prove to be unacceptably low, we have the board space and the architectural flexibility to attach two more cache chips to each processor. At the moment, we expect to achieve instruction/data cache balance (such that giving a new cache entry to either cache saves equal numbers of cycles) with twice as many entries for instructions as for data.

An important issue is how many Dragon processors an M-bus can support. The numbers above suggest that a single Dragon processor, with one cache chip for instructions and one for data, in steady state on a typical program, is likely to miss the cache about once every 13 machine cycles. Since a quadword read takes a minimum (and the average is probably quite near the minimum) of five machine cycles, the limit of system throughput is 2.6 processors' worth, independent of how many processors are available. According to Smith and Goodman's data, this limit would increase to 7.2 if a second cache chip for instructions were added to each processor. This is the main argument for increasing the size of Dragon's caches.

3.2 Virtual-To-Real Mapping

As mentioned above, addresses on the M-bus are real. Caches can intercept and implement many of the virtual-to-real mapping operations required by their processors. But what happens when they can't?

When it needs to know map information, a cache does an IORead from an address that includes the virtual page number to be mapped. (This is one reason why we need a 32-bit I/O address space.) The map processor responds to the IORead by providing the map information. While formulating its response, the map processor may do with the M-bus as it pleases, as long as IOReadDone is never asserted until the response is ready.

This means that the real memory itself can hold the map table entries for those pages of virtual memory that are present in real memory. One idea is to organize the resident map as a hash table occupying a constant small fraction of real memory. There would be an entry in the map corresponding to each virtual page present in real memory. The hash function would probably be the low-order bits of virtual page number. Collisions would be resolved by rooting a balanced tree of map entries, keyed by virtual page number, at the site of a collision in the hash table. Other nodes of the balanced tree would be stored in otherwise empty locations in the hash table. This yields an expected number of probes arbitrarily close to 1, depending on how big and empty the hash table is. It also yields log (virtual page count/hash table size) probes in the worst case. Other ideas are also under consideration, and the prototype machine will probably have a map processor that implements the identity function.

It is sometimes necessary to notify all caches that the map has changed, and ChangeFlags is a special M-bus transaction for this purpose. ChangeFlags behaves like an IOWrite with no data that requires no acknowledgment. The most common use of ChangeFlags is to break the virtual/real association of a real page in all caches. This is done just before writing a page to disk, for example, to prevent processors from changing data as it is being written.

Suzuki's measurements indicate that about 3% of the data references of a typical program require a map transaction on the M-bus. If the map processor is designed to cache any significant fraction of recently active map entries, it will respond to most inquiries in two cycles. From this, and the average of 7 machine cycles per memory operation, we expect the average cost of mapping to be less than 1% in execution speed.

4 DRAGON PROCESSORS

In previous sections we have dealt with the M-bus that binds Dragon together, and with the caches that attach to the M-bus. In this section, we deal with the processors, seen in figure 1, that attach to the caches. One of these is the I/O processor, a commercial microprocessor that interfaces the M-bus to a VME bus (an industry-standard 32-bit microcomputer bus) for dealing with purchased I/O controllers. One is the mapping processor, and one is the display controller. The remaining ones are identical Dragon processors, and they are the topic of this section.

4.1 Dragon Instruction Set

The Dragon processor has a new instruction set. This instruction set had to be considerably reduced in semantic complexity from previous machines, so that we could implement it in a fast pipelined two-chip engine, and handle the resulting inter-instruction pipeline hazards. At the same time, we wanted not to lose the code density that Mesa had achieved.

The variable state of a Dragon processor can be viewed as consisting of the following registers:

PC   a 32-bit byte program counter.

R   a 128-word x 32-bit operand stack cache.

L   a 7-bit register that indexes R. L is the base of a window of sixteen registers in R that are easy to address.

S   a 7-bit register that indexes R. S is the top of the expression evaluation stack in R. Many instructions implicitly (or explicitly) address R[S-2]..R[S+1], and increment S by [-2..1].

SLimit a 7-bit register that is compared against S to know when the operand cache is getting close to full.

A   a 16-word x 32-bit group of easily-addressed global variables.

IStack a control stack cache up to 16 elements long containing saved PC and L values.

a few other special-purpose registers that contain field descriptors, quotient or product, faulted memory addresses, whether the Reschedule interrupt is enabled, etc.

Dragon instructions consist of a one-byte opcode and zero, one, two, or four bytes of additional operands. The opcodes now defined fall into several general categories:

Unconditional jumps or calls to { I | [S] }, returns

Conditional relative jumps, decided by comparing [S] against { I | [S] | [L] | [A] }

Memory reads

{ [S] | [L] | [A] } := [{ [S] | [L] | [A] } + I]↑

Memory writes

[{ [S] | [L] | [A] } + I]↑ := { [S] | [L] | [A] }

Densely-encoded operations that produce a new top-of-stack

[S] := { I | C | [L] | [S] op I | [S] op [S-1] }

Densely-encoded operations that copy top-of-stack to register window

[L] := [S]

General three-operand instructions with less density

{ [S] | [L] | [A] } := { [S] | [L] | [A] } op { [S] | [L] | [A] }

In this table, I stands for immediate data (of several lengths) within the instruction, and C is a group of 12 registers containing popular constants like 0 and -1. The notation { a | b } means "either alternative a or alternative b", [a] means the contents of register a, and a↑ means the value in memory location a.

The remaining opcodes are called XOP's and treated as procedure calls to a vector of runtime support routines. Such a round-trip procedure call costs four machine cycles, so it is reasonable to implement many complex operations by XOP's instead of by special control within the processor, except for the purgative effects of their implementations on the instruction cache.

Dragon's stack cache R is similar to the stack cache in the C machine proposed by Ditzel, *et al.*, (1) and reminiscent of (but more flexible, efficient, and complicated than) that in the RISC. The most interesting (as estimated by the compiler) parameters, local variables, temporaries of the several most-recently-called and still-active procedures of the executing process are kept within R. On overflow, the least recently used registers in R are flushed to memory, and on underflow, they are reloaded from memory. This overflow and underflow detection is automatic (using SLimit and IStack), and flushing and reloading is implemented by code in trap handlers.

A calling procedure can send parameters to a called procedure simply by pushing them on the stack. The first instruction of the called procedure loads L with S-(number of parameters-1), and the parameters become its first few local variables. The called procedure returns results to its calling procedure by storing them in its first few local variables. The Return instruction then loads S with L+(number of results-1), placing the results on top of the caller's stack.

IStack is a cache for control as R is for data. It is also flushed and refilled from memory via trap handlers, and its purpose is to accelerate Call and Return instructions.

Conditional jumps come in two forms, one predicted to jump, and one predicted not to jump. The processor initially behaves as if the conditional jump will happen in the predicted direction, and uses the pipeline backup hardware (that must be present to handle page faults) to recover from mis-predictions. Correctly-predicted control transfers of all kinds take one machine cycle, plus instruction buffer refill time. Mis-predicted conditional jumps cost four cycles plus refill time. This puts a real premium on correct prediction, an interesting challenge for compiler writers.

To check whether dense code was still possible in Dragon's instruction set, we translated the Cedar version of Puzzle by hand, trying to simulate average compiler technology. Here is the result, compared with similar results for other machines (32-bit addresses unless noted):

Code bytes	Machine	Language	Remarks
1214	Alto	Mesa (early)	16-bit address, no checks
1369	Lilith	Modula2	16-bit address, no checks
1534	Dorado	Cedar	Mixed 16/32-bit address, range checks
1938	Dragon	Cedar	Range checks
2080	VAX 11/780	C	Pointers instead of subscripts, no checks
2500	68000	C	No checks
2736	RISC	C	No checks
3592	MIPS	Pascal	Global optimization, range checks
4744	IBM 370	Pascal	

At this moment only about half of the 256 possible Dragon opcodes are defined; the rest are implemented as XOP's. As we further tune Dragon's instruction set we expect code density to improve.

4.2 Structure

A standard Dragon processor cluster consists of five or six large chips. Two of these chips are caches that mediate between the processor and the M-bus, one for instructions and one for data. One or two are floating-point accelerator chips. The remaining two chips form the heart of the Dragon fixed-point execution engine.

The two chips that constitute the fixed-point execution engine form a single logical entity, shown schematically in Figure 2. Instructions and data flow downward in this diagram. Dragon uses a two-phase clock; the heavy horizontal lines marked A indicate

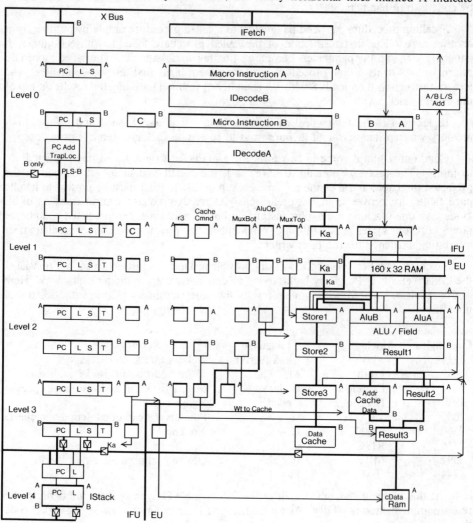

Dragon Processor Pipeline
Figure 2

latches that are open during the A phase, while those marked B show latches open during B. One of the chips is the Instruction Fetch Unit (IFU), which reads machine instructions, decodes them, expand them into microinstructions, and controls the pipelining of these microinstructions. The other chip is the execution unit (EU) that contains a 160-register 3-port data memory, a pipelined ALU/Masker/Shifter with hardware to assist multiply and divide, the address and data pathway to the data cache (but not its control), and several multiplexers to select operands and implement pipeline short-circuits. The heavy jagged diagonal line denotes the interface between the two chips. The line is shown as diagonal because as a microinstruction progresses through the pipeline, at each stage there is less work being done in the IFU and more being done in the EU for that instruction.

As we scan figure 2 from top to bottom, the first major block we encounter is IFetch. This block contains an asynchronously-filled 6-word buffer of instructions. When a whole instruction is ready, it is byte-aligned by IFetcher and made available to Level 0.

Pipeline Level 0 generates from each instruction a sequence (usually of length one) of fairly wide microinstructions. The registers holding L and S are at this level, and midway through this level the two fetch addresses for the EU RAM are formed.

Unconditional jumps, calls to literal addresses, and returns are completely implemented at Level 0. These operations take one machine cycle at that level, plus the time needed to refill the instruction buffer from the target address, which is one cycle in the best case.

In Level 1 the two EU RAM fetch addresses pass to the EU, and somewhat later literal data from the instruction, and information about the source of the ALU operands, pass to the EU.

In Level 2 the ALU operation and information about what Boolean condition is of interest passes to the EU, and somewhat later the EU does the ALU operation and computes the Boolean condition.

In Level 3 a data cache operation is initiated if appropriate. The address came out of the ALU in the previous level. The wide data path from EU to IFU at the end of Level 3 can carry new contents for an IFU register, such as the program counter. We chose to put data cache operations in a pipeline stage following the ALU, rather than in the same pipeline stage, even though the pipeline is longer that way. We did this because a very large fraction of memory instructions use an address formed as "base register plus offset", so the cache operation is preceded by address arithmetic. If ALU and data cache operations were in in the same pipeline stage, such memory instructions would require an extra cycle.

A microinstruction can fail due to a page fault, an arithmetic overflow, or the realization that a conditional jump has been mis-predicted. Between Level 3 and Level 4, a crucial decision is made. Only by this point is it certain whether a microinstruction will complete successfully. If it cannot, then not only that microinstruction, but all previous microinstructions for the instruction that generated it, as well as all microinstructions for following instructions, must be aborted. The processor must attain a state as if all microinstructions for previous instructions had successfully exited the pipeline (which they have), and all microinstructions for the instruction currently at Level 3-4 had never entered the pipeline (a condition, alas, quite contrary to fact).

In Level 4, successful instructions store new data back in the EU RAM (see Level 1), and successful calls or returns are recorded in the IStack.

4.3 Pipelining

Dragon derives much of its speed from its IFU and EU pipelining, and the fact that average instructions (in Puzzle, for example) operate with a pipeline headway of less than two cycles, even though the pipeline is four cycles long. Unfortunately this speed is not free. We must detect and resolve inter-stage dependencies (hazards). In addition, to deal with faults in data cache references, arithmetic faults of various kinds, and mis-predicted conditional jumps, we need a way to abort instructions. This involves recovering much of the processor state to its value three pipeline steps ago.

4.3.1 Hazards. Our design deals with inter-microinstruction pipeline hazards in three ways. The simplest is to avoid them. MIPS avoids hazards by a set of programming rules that are followed by the compiler. With this solution comes the choice of a very smart compiler or a lot of extraneous No-Op's. We didn't want to introduce such rules (and their attendant increase in code size) at the machine instruction level. We do avoid some hazards by implementing a few infrequently-executed instructions with awful side effects (Store IFU Register, for example) by a carefully coded sequence of microinstructions that clears microinstructions for preceding instructions out of sensitive places in the pipeline before the killer microinstruction does its dirty deed, and then prevents microinstructions for succeeding instructions from entering sensitive places in the pipeline while the deed is still in progress.

Another case of hazard avoidance is delay-matching in parallel paths. Dragon has a simple loop-free pipeline in which nearly every computational resource (gate, bus, etc.) is used at only one pipeline stage. This greatly simplifies pipeline scheduling, but it also means that some paths through the pipeline are longer than they would otherwise need to be. For example, consider the store-back path to the EU RAM. For ordinary arithmetic, this could be located at Level 3. But because we put the ALU operations and data cache references in tandem, rather than in parallel, for data cache references the EU RAM store-back path has to be at Level 4. A hazard is avoided by inserting an extra pipeline stage between ALU output and EU RAM store-back path, so that it is always used at Level 4.

Another way Dragon deals with hazards is to detect them, and to prevent some upper endangered segment of the pipeline from advancing, while allowing the lower danger-free segment to advance normally. A microcode No-Op, which changes no state and participates in no hazards, is inserted in the pipeline gap thus created. The boundary between the stationary initial pipeline segment and the advancing final pipeline segment is chosen as high in the pipeline as possible, but so that at least one of the two participants in any hazard is on the stationary side of the boundary. In Dragon, there are three such boundaries. One is at Level 0, and deals with next instruction not ready, or a Call-Return hazard. One is at Level 1, and deals with those cases of EU RAM Write-Read hazard that cannot be short-circuited. And one is at Stage 3, and deals with data cache not ready.

The final way Dragon deals with hazards is to detect them, and to achieve the desired effect in another way, without introducing delay. The data path round-trip, from the EU RAM, through the ALU, through a best-case data cache operation (or the parallel collision-avoiding pipeline stage), and back into the EU RAM, takes three pipeline stages. During that time a potential Write-Read hazard exists on the EU register that is specified as the destination of a microinstruction, because a following microinstruction can read the wrong value. Many of these hazards can be handled without wasted cycles by means of EU pipeline short-circuits, which are additional data pathways in the EU that carry data backward one or two pipeline stages. For example, there is a short-circuit from the output of the ALU back to either input, so that an instruction may use the arithmetic result of the previous instruction. The use of these short-circuit paths is governed by the IFU. It matches the source register addresses of an instruction about to enter pipeline stage 1 against the destination register addresses of instructions further down the pipeline. Short-circuit paths are expensive, both in area and complexity, so we have included them only where they will speed up common instruction sequences.

4.3.2 Instruction Abort. Dragon has a way of attaining a state as if all microinstructions beyond pipeline stage 3 ran to completion, while all those at stage 3 and earlier never started. This mechanism, combined with careful microcoding of multi-microinstruction instructions, allows a data cache fault, or an arithmetic fault, to behave as if an XOP (extended op, or very short procedure call) had been inserted before the instruction that caused the fault, and allows the instruction to be re-executed after the fault has been remedied. This same mechanism is also used to speed up correctly-predicted conditional jumps.

This state restoration is probably the one mechanism responsible for the most logical complexity in the processor. The mechanism is implemented in different ways for the different state variables involved. To recover state in the EU, where permanent state changes only take place at pipeline stage 2 and beyond, the IFU converts destination register addresses to an unimplemented register number, converts the ALU operation in Level 1 to a harmless OR, and notifies the ALU to recover Carry and other small amounts of state from internal saved versions. Within the IFU, the current program counter, S, and L registers at Level 1 are reloaded from shadow copies at Level 3.

The worst problem is the IFU's control stack cache, IStack. This cache contains the caller's return PC and L values for the most recent few pending procedure calls, XOP's, or Reschedule interrupts, or fault traps in the current process. To avoid a complex recovery process, this cache is located at Level 4, which is attained only by microinstructions that do not fault. Unfortunately, this means that if the matching Return appears fewer than three pipeline stages after a Call, the control stack cache does not yet contain the caller's return PC, so a Call-Return hazard exists and an interlock must be activated.

4.4 Multiprocess Primitives

Dragon will execute a distributed process scheduling algorithm. This algorithm is implemented by two hardware features and some software. One hardware feature is a global interrupt signal called Reschedule that is wired to all processors in a system. The other is a Dragon instruction called Conditional Store that implements a particularly nice form of atomic memory update:

$$\{[S+1] := [S-1]\text{↑}; \text{ IF } [S+1]=[S] \text{ THEN } [S-1]\text{↑} := [S-2]; S := S+1\}.$$

The scheduling algorithm is triggered when Reschedule is asserted, either by a Dragon processor or by the I/O processor. When that happens, each processor waits until the code it is running is willing to accept a process switch (handlers for stack overflow/underflow are not, for example), and then does a procedure call to the Reschedule handler code. That code attempts to acquire the global scheduling lock. If that fails, the code checks the processor's control block to ensure that this processor is still executing the process that the scheduler had in mind, and then returns. If the handler acquires the scheduling lock, it unloads the current process context to memory, loads the context of the scheduler process, runs the scheduler process (reassigning processes to processors), activates the Reschedule line once more while holding the scheduling lock, then unloads the scheduler's context to memory, releases the lock, and finally picks up and runs the process in the processor's control block.

The Dragon data and control caches only work well if processors do not switch processes too often. One way of avoiding process switches is to cause the Reschedule handler to reach for the scheduling lock somewhat faster if the processor is running the idle process than if it is not. This means that if there are any idle processors, one of them become the scheduler, and the others will only switch processes if forced by the scheduler code.

5 CEDAR

The designers of Dragon had the Cedar language uppermost in mind as the language from which Dragon programs would be compiled. Much of the organization of Dragon is intended to optimize the execution of common Cedar constructs. Of course, similar constructs can be found in other languages as well. Hill argues that the Bell Labs C machine, with which Dragon shares many features, is especially well suited to executing a wide variety of languages (4).

5.1 Procedures and Local Frames

Cedar is a successor of Pascal. It depends heavily on the notion that abstraction is implemented by procedure call. A process can be viewed as a LIFO stack of suspended activation records and one current activation record. Each activation record (or "frame") typically contains a program counter and some variables. To call a procedure, the currently-active procedure generates a new activation record for the called procedure, computes some arguments, stores them in the new activation record, suspends its own activation record, and makes the new activation record current. When a current activation finishes, it sends a number of results to the (suspended) caller's activation record, destroys its own activation record, and makes the caller's activation record current.

It is this computational paradigm for which Dragon is heavily optimized. Within the IFU chip is a LIFO cache of program counters and register indexes within the EU chip where the variables of suspended activation records begin. Most activation record variables can be implemented as EU registers, and parameter and result passing is done by index adjustment rather than data copying.

Such other languages as Lisp and Smalltalk also fit this computational paradigm more or less, and Dragon should be an effective engine for them as well. But there are other computational paradigms, such as Prolog, for which an equivalent amount of hardware, organized differently, would probably do significantly better than Dragon.

5.2 Strong Typing and "Safety"

Cedar is a strongly-typed language like Pascal. This means that the compiler can know a great deal about what might happen during execution, and what cannot. Cedar also has a garbage collector, like Lisp. This means that it must be possible, at run-time, to follow all pointers, and to know the syntactic types of their targets. In "safe" Cedar (an automatically-checked subset of the language) the use of the general "Address-Of" operator is prohibited, as are overlaid variant records, and other devices that can confuse the compiler about the semantics of execution. The result is that the compiler knows many situations when a variable in an active process does not need a memory address. Such variables can be implemented in machine registers without fear that a program might reference them by memory address "alias" and get an inconsistent copy.

A compiler cannot know in general whether alias references will happen in weakly-typed languages like BCPL or C. One solution proposed by Patterson is to simulate such alias references by loading or storing the appropriate register (6). If needed, Dragon will implement this solution by using the mapping hardware to trap ranges of memory addresses. This implementation would be unacceptably slow if such alias references were very common.

This is one illustration of the general observation that a smart compiler working with a strongly-typed language can do well if it has at least two coding templates available: one that is very efficient for common situations that it understands perfectly, and another one that always works but may be less efficient, for uncommon situations that it cannot predict with certainty.

5.3 Modules and Procedure Calls

Cedar, and its precursor Mesa, have very rich control structures including independent compile units, coroutines, and procedure variables. In our earlier Mesa instruction set we had just one kind of cross-compile-unit procedure call instruction, EXTERNALCALL, which favored compactness over speed. Every cross-module procedure call required at least four indirect memory references, accompanied by considerable unpacking and indexing, just to find the new program counter. Every callee needed a record in memory for his local variables, and that record was allocated from a special-purpose allocator, at a typical cost of two memory references. The benefit was that an external procedure call without arguments could be coded in just one or two bytes; the drawback was that where they needed performance, programmers would avoid using EXTERNALCALL by expanding procedures inline, thereby defeating EXTERNALCALL's dense coding.

In Dragon we relax the density of coding of EXTERNALCALL, and we permit varying degrees of indirection, corresponding to tightness of binding, as circumstances permit. A general Dragon EXTERNALCALL is five bytes long. In those five bytes the EXTERNAL-CALL can be represented in at least three different ways. Tightest of all is a simple procedure call to a 32-bit byte address. This requires no indirect memory references to find the new program counter, but it puts considerable pressure on the loader. Loosest of all is an XOP with a four-byte argument. The XOP interpreter might treat the argument as a pointer to a string variable containing the name of the procedure to be called, locate the appropriate routine via the loader tables, and complete the call. Somewhere in between is a three-byte data reference that loads a new top of data stack, followed by a two byte procedure call indirect through the stack top.

5.4 Processes

Cedar already contains a formally complete set of primitives for dealing with multiple tasks in a single address space: PROCESS, FORK, JOIN, WAIT, NOTIFY, MONITOR. We have considerable experience using these primitives in contexts where multiple processes seem natural. These will suffice as Dragon's multiprocess primitives until something better evolves.

6 CONCLUSIONS

We have presented the current state of the design of a high-performance general-purpose personal computer system for use in programming research. We are optimistic that its performance will be quite good in a single-processor configuration, and depending on the tasks at hand and the cache sizes used, modest numbers of processors interacting in the same virtual address space may not seriously impede each other. Further conclusions await further experience with the machine.

7 ACKNOWLEDGMENTS

Dragon was conceived in 1981 by C. Thacker, B. Lampson, P. Petit, N. Wilhelm, and F. Baskett. Today's architects and designers include R. Atkinson, R. Barth, D. Curry, J. Gasbarro, L. Monier, M. Overton, and myself. When the final story is told, there will be many others.

Good IC design tools from PARC's Computer Science Laboratory, and a cooperative spirit and functional silicon from PARC's Integrated Circuit Laboratory, have been and will continue to be essential in whatever success we may achieve.

Figure 2 was provided by D. Curry.

8 BIBLIOGRAPHY

1. Ditzel, D. and H. R. McLellan. Register Allocation for Free: The C Machine Stack Cache. ACM SIGPLAN Notices, XVII(4): pp. 48-56, 1982.
2. Goodman, James R. Using Cache Memory to Reduce Processor-Memory Traffic. Computer Architecture Symposium Proceedings, IEEE, pp. 124-131, 1983.
3. Hennessy, John, with Norman Jouppi, Steven Przybylski, Christopher Rowen, and Thomas Gross. Design of a High-Performance VLSI Processor. Third Caltech Conference on Very Large Scale Integration, pp. 33-54. Edited by Randal Bryant, Computer Science Press, 1983.
   Describes the Stanford MIPS machine.
4. Hill, Dwight D. An Analysis of C Machine Support for Other Block-Structured Languages. Computer Architecture News, published by ACM Special Interest Group on Computer Architecture, XI(4): pp. 6-16, 1982.
5. Kogge, Peter M. The Architecture of Pipelined Machines. McGraw-Hill, 1981.
6. Patterson, David A. and Carlo H. Sequin. A VLSI RISC. Computer, IEEE, XV(9): pp. 8-21, September, 1982.

7. Patterson, David A., with Phil Garrison, Mark Hill, Dimitris Lioupis, Chris Nyberg, Tim Sippel, and Korbin Van Dyke. Architecture of a VLSI Instruction Cache for a RISC. Computer Architecture Symposium Proceedings, IEEE, pp. 108-116, 1983.

8. Radin, George. The 801 Minicomputer. IBM Journal of Research and Development, XXVII(3): pp. 237-246, 1983.

9. Sites, Richard. How to Use 1000 Registers. Caltech Conference on VLSI, pp. 527-532. Edited by Charles Seitz, 1983.

10. Smith, Alan Jay. Cache Memories. ACM Computing Surveys, XIV(3): pp. 473-530, 1982.

11. Smith, James E. and James R. Goodman. A Study of Instruction Cache Organizations and Replacement Policies. Computer Architecture Symposium Proceedings, IEEE, pp. 132-137, 1983.

THE VLSI VAX CHIP SET MICROARCHITECTURE

Will Sherwood
Digital Equipment Corporation
Hudson, Massachusetts USA

ABSTRACT

VLSI technology has made it possible to compress the full functionality of the VAX-11/780 superminicomputer, at comparable performance, onto a 771,000-transistor microprocessor chip set. The eight-component chip set implements the full 304-instruction VAX computer architecture, 17 data types, and 4-Gbyte virtual memory management. Using a 40-bit microinstruction and a simplified 32-bit datapath, this VLSI implementation can execute frequently-used instructions in less than or the same number of microcycles as previous implementations, and has the same 200ns clock cycle time as the VAX-11/780. The chips are fabricated with a 3u (drawn) two-metal-layer NMOS process.

INDEX TERMS: VLSI, VAX Computer Architecture. Microarchitecture, VLSI CAD Tools

CHIP SET OVERVIEW AND PARTITIONING

The full VAX computer architecture [Levy 80, VAX 81a] is implemented in the VLSI VAX chip set microarchitecture [Johnson 84a,b]. The microarchitecture is partitioned as shown in Figure 1 into the instruction fetch

The following are Digital Equipment Corporation registered trademarks: VAX, VMS, UNIBUS

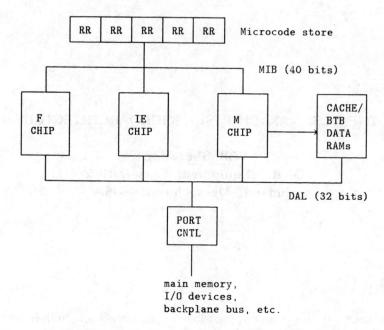

Figure 1a. Block Diagram - The VLSI chip set is driven by external microcode and is interconnected by two time-multiplexed busses time-multiplexed busses

Chip	Type	Transistor Sites	Package Pins	Signal Pins +
ROM/RAM Chip	custom NMOS	206K	44	30
IE Chip	custom NMOS	65K	132	98
M Chip	custom NMOS	70K	132	111
F Chip	custom NMOS	54K	132	67
Port Controller	CMOS gate arrays	3 x 2000 gates	120	100
Cache/BTB data	2K x 8 RAMs	10 x 16K bits	24	22

+ Other package pins are power/ground/unused.

Figure 1b. Chip set summary

and execute (IE) chip, the memory mapping (M) chip, and the optional floating-point accelerator (F) chip. The IE chip contains the main 32-bit datapath, hardware to prefetch, decode, and execute VAX instructions, and a small five-entry translation buffer, the mini-TB (MTB). The M chip contains the tags and sequencer for an 8K-byte cache, the tags and sequencer for a 512-entry backup translation buffer (BTB), interrupt logic, timers, and 4 UARTs. The F chip contains a 67-bit mantissa datapath and 13-bit exponent datapath for faster execution of most 32-bit and 64-bit floating-point instructions.

The IE, M, and F chips are connected via two major busses: the microinstruction bus (MIB), and data and address lines (DAL). These busses are time-multiplexed in two half-cycles. The 40-bit MIB carries a microaddress plus various inter-chip status signals during the first half of a cycle and a microinstruction during the other half. Similarly, the 32-bit DAL carries a memory address during the first half of a cycle and memory data during the other half. The DAL is connected to main memory, I/O devices, and a backplane bus via three gate arrays which make up the port controller logic.

The chip set's microcode is contained in an off-chip control store hybrid, which consists of five custom 100 ns-access ROM/RAM chips. Each ROM/ RAM chip contains 16Kx8 of ROM, 1Kx8 of patch RAM, and 32 words of CAM (content addressable memory for identifying patched ROM locations). The five chips make up 16K of 40-bit words of microcode, 1K of 40-bit words of patches, and 160 CAM locations.

The chip set requires minimum support circuitry. Thus, clock generators, address latches, interrupt logic, timers, and console support are included in the chip set. This reduces the number of pins devoted to external busses. The only external chips provide power, clocks, cache and BTB RAMs, and RS-232 driver/receivers for the console terminal.

The chip set is compared with other VAX family members in Figure 2.

	VLSI VAX	VAX-11/780	VAX-11/750	VAX-11/730
relative performance (VAX-11/780 = 1.0)	0.9	1.0	0.6	0.3
microcycle time (ns)	200	200	320	270
cache : size type block size allocate size	8k bytes direct mapped write-thru 16 bytes 64 bytes	8k bytes 2-way set associative write-thru 8 bytes 8 bytes	4k bytes direct mapped write-thru 8 bytes 8 bytes	none
translation buffer allocate size	2 levels 5 entries + 512 entries direct mapped 4 entries	1 level 128 entries 2-way set associative 1 entry	1 level 512 entries 2-way set associative 1 entry	1 level 128 entries direct mapped 1 entry
execution time (microseconds) for ADDL2 reg, reg	0.40	0.40	0.93	2.72
microstore size	16k ROM 1k RAM	4k ROM 2k RAM	6k PROM 1k RAM	16k RAM
microword width(bits) (parity bits)	39 1	96 3	78 2	23 1

Figure 2. Comparisons of the chip set with other VAX family members

MICROINSTRUCTION FORMAT

The microinstruction format is heavily encoded horizontally in both the micro-operation and branch fields. Because of the delays associated with pipelined conditional microbranches, a two-level vertical/horizontal micro-code structure is used. With careful encoding and multiple instruction formats, a single 40-bit microinstruction describes almost as much work per cycle as the 80- and 99-bit microinstructions used in the VAX-11/750 and VAX-11/780, respectively. The execution of each microinstruction overlaps with the fetching of the next one.

As shown in Figure 3, each microword contains fields that can specify the following four concurrent hardware activities:

- o ALU, shift, memory, or special purpose function with source and destination
- o Condition code manipulation
- o Miscellaneous hardware function
- o Microsequencing branch

Every microinstruction contains a 1- to 7-bit format (micro-opcode) field that specifies the class of operation to be performed. The associated function microfield contains the intended function for the selected hardware. For example, the function field for an ALU format microinstruction can specify add, subtract, or logical operations.

Typical microinstruction formats control passing data from the two EBOX source busses to the destination bus. The source/destination field specifies which register sets are driven onto the two source busses and which register set is written from the destination bus. The A and B fields are then interpreted as the register address within the register set for the particular source and destination. All of the above information is packed into 18 bits. In a fully-expanded encoding scheme, 63 bits would be needed.

The chip set supports two sets of condition codes: the VAX condition codes (PSL CCs), which have different meanings (and are set differently) depending on the current macro opcode, and the "raw" ALU condition codes (ALU CCs), which are set for each ALU operation. Both sets of condition codes can be altered every microcycle in parallel with other hardware activities. The ALU CCs can be set according to various data lengths, and the PSL CCs can be set from the ALU CCs (or a mapping of them as selected by the current macro opcode). Condition codes and

Basic ALU Format

38	37	36	35	34	33	32	31	30	29	28	27	26	25	24	23	22	21	20	19	18	17	16	15	14
0	ALU Function			Src/Dest			CC		B Reg			MISC				A Reg				

Constant/ALU Format

38	37	36	35	34	33	32	31	30	29	28	27	26	25	24	23	22	21	20	19	18	17	16	15	14
1	0	ALU		Formt	Src/Dest	Pos'n	CC	Constant Data Byte								A Reg				

Shift Format

38	37	36	35	34	33	32	31	30	29	28	27	26	25	24	23	22	21	20	19	18	17	16	15	14
1	1	0	Shift Amount			Src/Dest			CC	B Reg			MISC				A Reg				

MXPR Format

38	37	36	35	34	33	32	31	30	29	28	27	26	25	24	23	22	21	20	19	18	17	16	15	14
1	1	1	0	Reg Addr				RD	CC	Src/Dest		MISC				A Reg				

Memory Request Format

38	37	36	35	34	33	32	31	30	29	28	27	26	25	24	23	22	21	20	19	18	17	16	15	14
1	1	1	1	0	0	Mem Func			FF	CC	Src/Dest		MISC				A Reg				

FBOX Execute and Transfer Formats

38	37	36	35	34	33	32	31	30	29	28	27	26	25	24	23	22	21	20	19	18	17	16	15	14
1	1	1	1	0	1	0	F BOX OP		CC	Src/Dest		MISC				A Reg				
1	1	1	1	0	1	1	FBOX MISC														

Special Format

38	37	36	35	34	33	32	31	30	29	28	27	26	25	24	23	22	21	20	19	18	17	16	15	14
1	1	1	1	1	0	MISC 1			MISC 2			MISC 3			A Reg				

Spare Format

38	37	36	35	34	33	32	31	30	29	28	27	26	25	24	23	22	21	20	19	18	17	16	15	14
1	1	1	1	1	1	Function			CC	B Reg			MISC				A Reg				

Microsequencer Control

13	12	11	10	9	8	7	6	5	4	3	2	1	0	
0	BRANCH COND. SEL			BRANCH OFFSET										Branch Format
1	SB	JUMP ADDRESS												Jump Format

Figure 3. Microword Formats

data length are encoded in 2 bits of the microword.

The miscellaneous microfield allows an unrelated piece of hardware to be utilized on every cycle. Choices include controlling the I-stream prefetcher, writing an additional register destination in parallel with the specified destination, decrementing a counter, and setting data length. The miscellaneous field is encoded in 5 bits.

The microsequencer branch field provides a powerful set of conditional branches, unconditional branches, and 16-way branches. All the branching information fits in 14 bits, while producing an exceptionally rich (and hence fast) set of branches.

The remaining bit of the microword is a parity bit.

IE CHIP ORGANIZATION

The IE chip is organized as four separate subchips working in parallel (as shown in Figure 4). These four subchips can work on three separate simple VAX instructions or can work synergistically on a single VAX instruction.

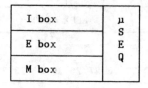

Figure 4. Subchips of the IE Chip

The memory interface subchip performs VAX address translation and protection checking and handles variable-width operands, unaligned operands, and page-crossing checks. The memory interface can start a new 32-bit memory reference every microcycle. The fastest read path (MTB hit and cache hit) takes one cycle. Similarly, the fastest write path (when an empty write buffer exists in the port controller) takes one IE chip cycle.

The E box subchip contains the main datapath. It is directly controlled by the microcode and can perform a 32-bit ALU or shift operation every microcycle.

The I box subchip prefetches, buffers, and decodes VAX instructions. Typically, it decodes one VAX instruction ahead of the execution and uses the memory interface to prefetch two VAX instructions. It can decode one to four bytes of instruction stream every microcycle.

The microsequencer subchip calculates the microcode next-address using branch and microtrap conditions from the other subchips every microcycle.

MEMORY INTERFACE SUBCHIP

The 4-Gbyte VAX architecture virtual address space is divided into sets of virtual addresses: those used by user processes and those used by the system. The VAX memory management scheme not only provides a large address space for instructions and data, but also allows data structures up to one billion bytes, and provides convenient and efficient sharing of instructions and data, across processes.

Three separate operations must be performed on each virtual address: checking for unaligned operands, checking for valid access (memory protection), and mapping to a physical address. All three of these operations are done by the memory interface of the IE chip.

To meet the needs of high-level programming languages, VAX data operands can be unaligned, e.g., a four-byte operand can start at a byte address that is not a multiple of four. In general, accessing an unaligned operand can cause multiple memory references, and the operand can cross two different pages. When needed, the memory interface hardware automatically performs both the two memory accesses and the associated masking and shifting for a single unaligned operand. This hardware sequencing gives higher performance than the microcoded solutions used in other machines. The typical unaligned operand is fetched from the cache in exactly two microcycles.

To implement these features, the memory interface section consists of a DAL sequencer, a mini-translation buffer, address registers, access and protection logic, and control logic.

The MTB holds the information needed to check access and translate addresses for five pages: one entry for instructions and four entries for data. If the virtual page address matches an MTB address tag, the access protection is checked, and the translated physical location is referenced, all in a single microcycle. This allows about 70% of all virtual memory references to be translated into physical addresses with no lost cycles. If the virtual page address is not found in the MTB, an MTB "miss" occurs.

The chip set has a unique second-level backup translation buffer (BTB) that is located on the M chip and in external RAM chips. This allows most MTB misses to be resolved in a single microcycle, with the original memory access retried on the second cycle (see Figure 5).

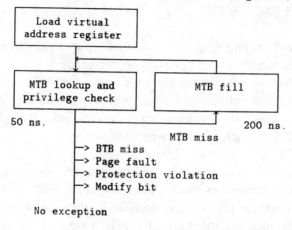

Figure 5. Translate Sequence, handled by hardware

In the infrequent case of an MTB/BTB miss, microcode accesses the third-level page tables in main memory in order to fill the MTB/BTB entry. The effect of the MTB/BTB pair is to have almost the performance of a 512-entry translation buffer in the silicon area of the 5-entry MTB.

E BOX SUBCHIP

The E box subchip allows for one-cycle execution of all 32-bit ALU and shifter operations. It uses standard dual precharged source buses and a single timeshared result bus. The scratchpad registers are dual-ported, allowing full read-op-write or read-op-test in one cycle. The user-visible general purpose register set (GPRs) and other architecturally defined registers are also located in the E box. E box microinstructions can initiate virtual or physical memory references and transfer the data to or from the scratchpad or architecturally-defined registers.

I BOX SUBCHIP

The I box subchip contains the instruction prefetch sequencer and associated datapath. This subchip is normally free-running in parallel with the other subchips. The I box can initiate virtual memory references to the instruction stream.

The I-stream prefetcher consists of an 8-byte queue that is filled from the macroinstruction stream in main memory or cache. If the queue is not full, it uses spare DAL cycles to read from cache via the memory interface subchip. As long as there are no memory management complications, the prefetcher runs in parallel with microinstruction execution. Because prefetched instructions are cached, main memory cycles are kept at a minimum.

The I box datapath provides hardware assistance for each of the several phases involved in decoding a VAX instruction. VAX instructions consist of a 1- or 2-byte opcode, followed by up to six operand specifiers. The operand specifiers are 1- to 17-bytes in length. Most operand specifiers contain an address mode, a register number, and possibly a 1- to 16-byte literal. The I box hardware can load up to four bytes of the instruction stream into dedicated registers during every microcycle.

The I box generates microcode dispatch addresses for the microsequencer subchip. The dispatch addresses are a function of the opcode and/or the current specifier's address mode. Special optimized dispatch addresses are generated for some register operands. The I box allows the decoding of most of the VAX instructions to be overlapped entirely with the previous instruction's execution.

MICROSEQUENCER SUBCHIP

The microsequencer subchip calculates the next microaddress for each microcycle. It contains a stack for microsubroutine call and return microbranches, and multiplexors and adders that implement the various conditional and relative branches in the microarchitecture.

The microsequencer maintains a microprogram counter (uPC) that points to the current microinstruction. The default next micro address is uPC + 1. Conditional branching is accomplished by adding a signed offset to the uPC to form the "true" branch microaddress and by taking the default uPC + 1 microaddress for a "false" branch. This uPC-relative scheme provides a high degree of flexibility to the microprogrammer, and the offset is encoded in only 7 bits.

Unconditional branches select the next address within a 4K block of microcode. When the subroutine bit in the branch field is set, the current uPC + 1 is pushed onto an 8-entry microstack. The return branch field adds the branch offset field to the address on the top of the microstack, thus allowing multiple return points. 16-way branches perform a logical OR of four bits of machine state with low-order address bits of the branch address. Depending on the pattern of low-order "1" bits in this address, a 2-, 4-, 8-, or 16-way branch can be achieved. Special broad jumps have an unlimited range, but cannot be used in the same cycle as an ALU operation. The net effect of the various branch and jump formats is to pack 14 bits of microcode address plus 6 bits of condition select into a total of only 14 bits.

Microtraps handle exceptional conditions such as parity errors, arithmetic overflow, or virtual address translation faults. Microtraps abort the current microinstruction, along with its prefetched successor. A microtrap then pushes the current uPC + 1 on the microstack, and the uPC is forced to a hardwired address. Execution proceeds with the first word of the exception routine at the forced address. Some exceptions eventually return to the address-minus-1 saved on the stack to retry the original microinstruction.

Microtraps allow hardware to watch for exceptional conditions, rather than wasting microcycles on explicit testing for rare events. This microsequencer design speeds up virtual memory mapping and simple arithmetic instructions.

M CHIP ORGANIZATION

The M chip contains a number of system support functions that contribute both to overall system performance and to small total chip count. The most important performance features are the backup translation buffer (BTB) and the cache support. The chip-count minimization features include the on-chip clock generators, timers, interrupt logic, and a UART for the console terminal.

Virtual references that miss in the MTB are tried in the BTB, under control of the M chip. The BTB consists of 128 virtual address tags stored in a dynamic RAM array within the M chip and 512 page table entries (PTEs) stored in an off-chip static RAM array. Each BTB tag is associated with four consecutive PTEs. Allocating four PTEs per tag saves valuable space in the M chip tag array while exploiting the probability of clustered virtual page references and minimizing keeping the miss ratio.

Having the BTB tags on the M chip allows a fast virtual address comparison and hence, only a single-cycle delay for most MTB misses. As an additional performance feature, the BTB hardware supports a software-initiated two-cycle invalidation of all process-space tags for fast context switching.

The M chip contains 128 cache tags that address 8K bytes of off-chip cache data RAM. For memory reads, the M chip performs a cache tag comparison against the physical address while the cache data RAMs fetch the corresponding 32-bit longword. If the cache tag matches, the reference completes in one cycle. When a cache miss occurs, the port controller completes the memory request.

The cache is filled 16 bytes at a time. This optimizes the bandwidth requirements of the main memory bus and often has the effect of prefetching bytes that are about to be used. The aligned 16-byte blocks are fetched as four 32-bit longwords. To reduce the time delay for a cache miss, the first longword fetched is the one that caused the miss, and the IE chip is only stalled until the data arrives on the DAL. The remaining three longwords needed to fill the cache come from main memory in wrap-around order (for example, 2-3-0-1, if the first longword had low-order address bits equal to 2) on the following three bus cycles. This design is an average of one and one-half cycles faster than fetching the longwords in a fixed order. Each cache tag is associated with four consecutive cache blocks (totaling 64 bytes).

The M chip also contains a number of loosely-related facilities that either increase system performance or reduce total chip count.

The address translation logic (ATL) on the M chip calculates the address of the required page table entry in memory after a BTB miss.

A PLA on the M chip compares the current program status longword (PSL) to a proposed new PSL. This logic is used during context switching to determine if the new PSL is valid.

The NMOS clock generators, VAX-defined interval timer, and time-of-day clock are also on the M chip.

Interrupt logic on the M chip funnels all interrupt requests to a single bit to be sent to the IE chip, and does appropriate masking and enabling of interrupts.

Finally, the M chip contains logic to allow microcode access to a console terminal device.

F CHIP ORGANIZATION

The optional F chip is a tightly-coupled floating-point accelerator for the IE chip, executing a subset of the VAX floating-point instructions. The IE chip executes all 128-bit and some very simple (move, compare) floating-point instructions, plus the remaining floating-point instructions (emulated in microcode) if the F chip is not present. The F chip architecture takes advantage of the possible parallel operation of the two chips by assigning the decision-intensive tasks to the IE chip while the F chip takes the burden of the actual calculation. The F chip runs internally at twice the clock rate of the rest of the chip set.

The F chip executes the following VAX instructions:

    ADD (F,D,G)   SUB (F,D,G)   CMP (F,D,G)   CVTL (F,D,G)
    POLY (F,D,G)   MUL (F,D,G,L) DIV (F,D,G,L)
    EDIV   EMUL

The F chip is synchronized with the IE chip by monitoring the MIB, which contains the VAX opcode, status, and microinstructions. At the beginning of each VAX instruction, the IE chip sends the new opcode to the F chip and asserts an initial instruction decode (IID) pin. The assertion of IID causes the F chip to suspend any current activity, reset all internal sequencers, and latch the VAX opcode. If an instruction is to be executed in the F chip, then the F chip uses this VAX opcode to determine number and data type of the operands. The F chip monitors the microinstruction stream looking for SYNC instructions that signal that F chip operand data is on the DAL. The F chip design allows memory operands to be latched by the F chip from the DAL as they appear directly from memory. This feature does not require the IE chip to waste cycles forwarding memory operands.

When the F chip has received the required number of operands, it starts the operation using on-chip microcode. While the F chip does the operation, the IE chip can check special cases or calculate the destination address. When the IE chip is ready for the result (IE chip stalls until the F chip result is ready) and the F chip has completed its execution, the IE chip reads the F chip's result register into an IE chip scratchpad or architecturally-defined register. The N and Z status lines from the F chip are loaded into the IE chip's condition code logic.

All floating point faults cause the IE chip microcode to microtrap and read the type of fault from an F chip register. The F chip function (ADD, MUL, etc.) can also be forced directly by microcode, which allows micro-diagnostics to run without executing a specific VAX instruction.

ROM/RAM CHIP

A traditional ROM-based microcode control store is expensive to update for microcode changes in the field. One alternative is a RAM-based control store, as used in the VAX-11/785 and other machines (see Figure 2). A drawback of using only RAM is that some non-volatile medium (such as a floppy disk, EEPROM, or down-line loading) could be required to hold the microcode so the RAM can be loaded each time the system is booted. This not only increases cost and boot time, but also can excludes using this system in a harsh, space-limited, or isolated environment. A second drawback of a RAM-based control store is that high-speed RAMs are not as dense and fast as ROMs.

The VLSI VAX chip set design implements the microcode control store in custom VLSI chips consisting of ROM, RAM, and content addressable memory (CAM) for patches [Calcagni 84]. This method employs the favorable speed and density characteristics of ROM while allowing patches for the ROM to be handled by RAM and CAM. This eliminates costly field ROM replacement when a microcode change is required.

As shown in Figure 6, each ROM/RAM chip consists of 16K 8-bit words of ROM, 1K 8-bit words of RAM, and 32 14-bit words of CAM. The chip set uses five of these chips in implementing 16K 40-bits words of ROM, 1K 40-bit words of RAM, and a total of 160 14-bit words of CAM.

In the actual operation, 14 address bits select a ROM location. and at the

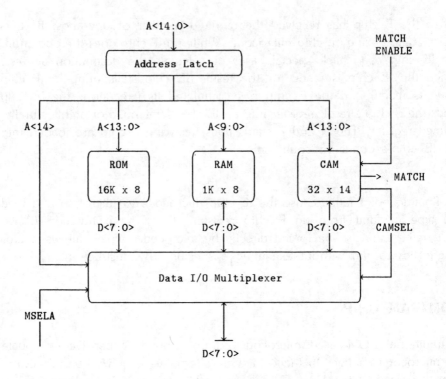

Figure 6. Block Diagram of the Patchable Control Store Chip

same time are compared against all 32 CAM locations within each chip. while the low-order 10 address bits select a RAM location. Normally none of the CAM locations match, and each chip's internal multiplexer selects the ROM output as the final data (Figure 7a). If a CAM match does occur, the MATCH line is fed back to the multiplexer to select the RAM output (Figure 7b). This multiplexer selection is done through a 15th address bit. For patch sequences consisting of more than one micro-instruction, the first instruction is patched via the CAM, and this instruction jumps directly to other instructions in the RAM (15th address bit asserted). The last instruction in the patch sequence jumps back to the ROM (15th address bit deasserted).

If it becomes necessary to modify one of the ROM locations in the field, it can be patched by programming the RAM and CAM, rather than changing all the ROM chip masks. In order to patch an erroneous microinstruction in ROM, the replacement instruction is written in the RAM location whose low-order 10 bits are the same as those of the ROM address, and its ROM address is loaded into one of the 160 CAM locations.

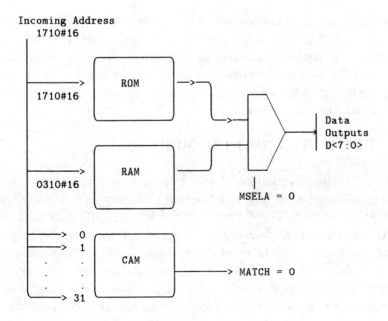

*Figure 7a.*Control Store Operation -
Valid ROM Microinstruction

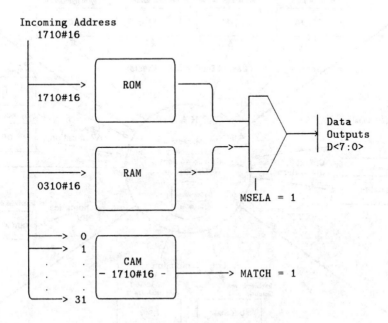

*Figure 7b.*Control Store Operation -
Patched Microinstruction

118

The patches normally are loaded at bootstrap time by microcode from a small, slow EEPROM. System software can load additional patches using privileged macroinstructions as part of the normal operating system bootstrap procedure. These additional patches can come from any supported I/O device, including disks and local networks. In addition, patches can be loaded using console commands.

CAD TOOL SUITE AND METHODOLOGY

To allow the implementation of the architectural complexity of this chip set, parallel development of a sophisticated CAD tool suite and methodology (Figure 8) was undertaken. A flexible, yet efficient, CAD data base manager (CHAS) interfaces both existing and new CAD tools to the logical and physical databases. CHAS provides a single point of departure for the designer's daily activities. Realizing the functionality for first-pass chips is based on the hierarchical simulation tools that link the architectural design to circuit design, as well as the sophisticated interconnect verification (IV) program that verify the correctness of the physical interconnect and permit the extraction of parasitic capacitances on critical path nodes.

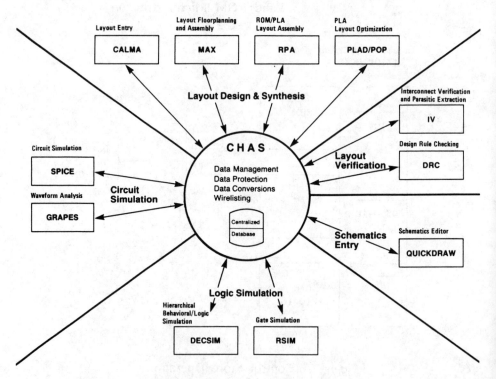

Figure 8. Suite of CAD Tools used in VLSI VAX Chip Set Design

Any VLSI design must have a longer up-front design time since the first-pass chips must strive for perfection; no prototype ECO correction wires are possible. Thus, the heavy use of CAD tools is a radical departure from the TTL breadboard prototypes. The total time-to-working-system is about the same for both the previous approach and the necessary VLSI approach. However, the ratio of design and verification time is far greater for VLSI (about 85%) than for a traditional TTL breadboard approach (about 50%).

Extensive software modeling from behavioral models of the system to behavioral models of chips, gate (logic) level models, and MOS circuits was an absolute requirement for building a machine of this complexity and performance. The broader the space the modeling effort covers, the shorter the development cycle, and fewer design change iterations occur from microcode to logic to circuit to layout.

The custom VLSI chips and microcode are developed in parallel. Instead of building a hardware prototype, software simulators are used for early verification of the design [Kearney 84]. A hierarchical simulation model description strategy was adopted because the requirements and cost tradeoffs were different at each step of the verification. Figure 9 shows the different levels of modeling used. The cost of simulation of a detailed transistor-level model for all the steps was prohibitive. The goal was to increase confidence in the model because it was used for hardware design verification. Architectural verification, described in [Gries 84], was accomplished using the high-level FUNSIM model. The high-level FUNSIM model was faster, but did not include some of the subtle implementation details in the corresponding lower level models. Therefore, tests were run first on the higher-level models. A test subset that maximized the test coverage was then run on the lower level models. The suite of models achieved maximum

Model Level	Simulator	Language/Style	speed+
FUNSIM	DECSIM	behavioral description/procedural	0.3
LOGSIM	DECSIM	detailed logic equations/procedural	1.6
TRSIM	RSIM	wire list/event driven	62.5

+ Speed was measured in CPU Seconds per microinstructions on a VAX-11/780.

Figure 9. Different Levels of Simulation

test coverage of the design with acceptable simulation time. Cross-checking and correlation between the models improved the trustworthiness of the high-level model.

Architecture validation and microcode verification were performed using an architecture validation program [Bhandarkar 82]. The validation program is an architectural exerciser that generates test programs to be run on both a host VAX and the chip set under test. The results are compared and scored. The exerciser and other diagnostic programs were run on the FUNSIM model, since it was fast and allowed many test cases to be run.

PROTOTYPE ENGINEERING TESTER

A custom test system was developed to integrate microcode and components of a VLSI VAX microcomputer prototype system [Sherwood 84]. Microcode debugging ease (on the hardware) and leverage is gained by hierarchically accumulating access and execution operations for the prototype. VAX-11/750-based software and custom UNIBUS-based hardware comprise this Engineering Tester (fondly called ET). The prototype system is controlled by interactively entered high-level commands to ET software (based on the command language for the DECSIM logic simulator [Sherwood 81]), which in turn communicates with the tester hardware and the prototype through UNIBUS control and data registers (Figure 10).

Once the chips were fabricated, they were individually tested and characterized on a Fairchild SENTRY-21 VLSI tester for functional verification and correct AC characteristics. The first-pass chips were 95% functional, a success largely due to the pre-fabrication CAD tool design verifications. The first-pass chips were immediately integrated into the prototype system controlled by ET. The same micro- and macrodiagnostic programs that were run on the simulator were then run on real hardware. This provided the engineers with a machine that was able to run VAX macrocode within two weeks, supplying valuable data for second-pass chips.

ET enabled the engineers to load and run micro- and macrocode, and had the full set of features that are typically only available in a system programmer's interactive debugger [VAX 81b] or an advanced logic simulation tool (see Figure 11). The high-level features of ET managed mundane tasks for the engineer, so the design problem could be investigated

UNIBUS

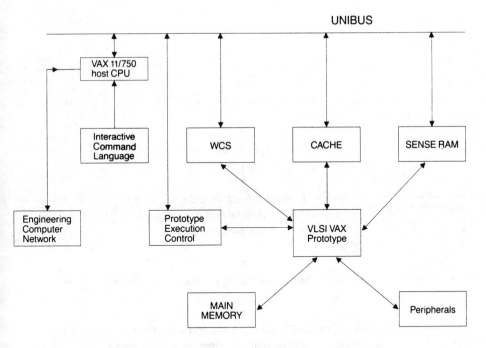

Figure 10. ET Block Diagram

with minimal overhead. It was a boon to have the power of a VAX (its computation ability, file system, line printer, connection to network) available for assembling, loading, searching, formatting, and storing tests and test results. When the symptoms of a problem were discovered. it was easy to set up the conditions for various experiments to isolate and pinpoint the source of the problem. An appropriate correction could then be made in microcode (and assembled and re-loaded), or, if it was a hardware problem, the corrected simulation model could be re-run to verify that the problem was fixed. In some cases, a hardware problem can be avoided by detecting the conditions associated with the problem using a WATCH command. This command intercepts microcode execution and then executes one or more additional commands to modify the system state and superimpose the corrected result. In either case, having these bugs corrected allowed the continuation of further testing so that other problems could be discovered. These microcode corrections and command file patches were often collected into indirect command files and then run automatically so the users can assume correct chip functions and continue with debugging. Through the use of ET, the first-pass chips were patched to be almost 100% functional.

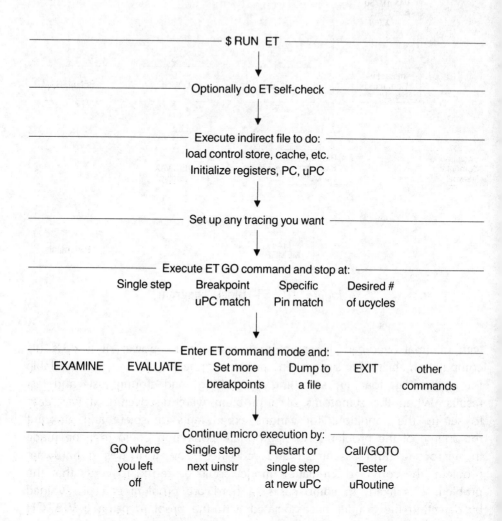

——————————————— $ RUN ET ———————————————

↓

——————————— Optionally do ET self-check ———————————

↓

——————————— Execute indirect file to do: ———————————
load control store, cache, etc.
Initialize registers, PC, uPC

↓

——————————— Set up any tracing you want ———————————

↓

——————————— Execute ET GO command and stop at: ———————————
Single step   Breakpoint   Specific   Desired #
                   uPC match   Pin match   of ucycles

↓

——————————— Enter ET command mode and: ———————————
EXAMINE   EVALUATE   Set more   Dump to   EXIT   other
                               breakpoints   a file   commands

↓

——————————— Continue micro execution by: ———————————
GO where   Single step   Restart or   Call/GOTO
you left   next uinstr   single step   Tester
off   at new uPC   uRoutine

Figure 11. Typical Commands Used in ET Session

RENDERING THE VAX ARCHITECTURE IN VLSI

The first implementation of the VAX computer architecture in the VAX-11/780 was designed using TTL and printed circuit board technologies. Many of the architectural design decisions were made with TTL implementations in mind. Some architectural aspects did not fit easily into VLSI, but architectural compatability could not be compromised in this VLSI implementation. This along with performance goals constrained many aspects of the microarchitecture and associated microcode. The following sections discuss some of the benefits and drawbacks of implementing an already existing architecture in VLSI.

DIFFICULTIES IN ADAPTING FOR VLSI

In comparing the VLSI VAX chip set microarchitecture to that of the VAX-11/780, we found the following areas to be more difficult to design with a good engineering balance:

Bit field alignment and rotation is free in TTL/PCB designs: the wires are simply re-routed in the desired permutation. Several examples of bit positioning trade-offs in the VLSI VAX were:

PTEs were rotated to get the page table entry address information aligned to perform the virtual address translation arithmetic.

The CALLS procedure call instruction packs some status bits into a 32-bit longword to be pushed on the stack. In TTL, the bit packing/rearrangement is trivial; in VLSI, it was deferred to microinstructions that mask, shift, and OR the bits into position.

The most frequently accessed macro privileged register is the interrupt priority level (IPL) register. This register is a field in the program status longword (PSL$<20:16>$). It was too expensive to have the hardware align the register data to be right-justified in the least significant bit, thus the microcode performs the shift.

Some floating point field extraction was deferred to microcode.

Translation buffer chip area is a little large (wide) with respect to the 512-byte page size.

Selective byte and word writes to the GPRs increased the write control logic size and complexity.

Maintaining residual register data for the string and decimal macro-instructions use microcode temporary registers, microcycles, and microwords without doing useful work.

The number of temporary registers in VLSI is relatively expensive. In TTL, 256-by-4 register chips, for example, can be used.

Pin limitations for VLSI chips were constraining in two major areas:

It restricted the width of the microword to 40 bits

The interface between the IE chip and F chip as well as the organization of the F chip were severely constrained. The VAX-11/780 dedicated 106 wires between the main processor and the floating point boards, and there were GPR shadow copies in the floating point accelerator for easy and quick operand access. The VLSI trade-off was to have the F chip "listen in" on operand fetching and make explicit transfers of GPR data operands.

WHERE THE DESIGN BENEFITED FROM VLSI

Often less logic can be expended for a function in VLSI because the circuitry is custom and not constrained to off-the-shelf gate configurations or PLA/ROM sizes. The IBOX specifier decoding PLA had fewer bits than the VAX-11/750 because the 750 used fully expanded ROMs (512 instruction possibilities) and had all six possible specifiers included. The VLSI implementation employed a PLA having only the needed 304 instruction terms and only 2 specifiers (the other specifiers, used in infrequent instructions, were handled explicitly in microcode).

Invalidating the translation buffer (for instance, to change process contexts) was easier in VLSI. A single line run through the RAM array cleared the single valid bit in each entry. In TTL, each RAM location had to be read and masked separately (taking many cycles).

The on-chip mini-translation buffer eliminated on- and off-chip delays and was more fully integrated into surrounding logic.

The EBOX subchip was able to provide more custom functions:

Use branch conditions to detect data values greater than 31 (in addition to the conventional sign-bit conditional branches)

Test each byte of a 32-bit value for equal-to-zero

Use decimal addition correction logic to adjust by 6 or 3

Even though there were trade-offs made in implementing the VAX computer architecture in VLSI, the benefits of the resulting small form factor at comparable speed far outweighs the few functions that were microcoded in lieu of dedicated hardware.

CONCLUSIONS

Rendering the complete VAX architecture in VLSI confirmed the approach that the designer should heavily critique the defined microarchitecture before beginning MOS design. This is due to the enormous expansion factor that results from the complexity of translating the microarchitectural functions and wires into transistors and silicon area. If the microarchitecture is complicated and complex, the silicon level design becomes far more complicated. The designer should make the microarchitecture as simple and clean as possible, which then gains much design time and fewer bugs in the silicon design. This point may seem obvious; however, the review of a microarchitecture design from the standpoint of making it simpler cannot be overemphasized. Every bit in the microword should be examined as well as every function. The payoff in simplifying designs and eliminating functions and wires at the beginning of a project is tremendous for a smooth silicon design process. In order to do this well, the designer must have an absolutely solid and clear understanding of the hardware's architectural usage by programmers, so that effective tradeoffs can be made during microarchitectural design reviews.

ACKNOWLEDGEMENTS

The author wishes to thank the following people for their help and support for this chapter: Duane Dickhut, Dick Sites, Vince Pitruzzela, Suzanne Laforge, Cliff Farmer, Laura Yuricek, Bob Gries, Rick Calcagni, John Brown, Charles Lo, Sri Samudrala, Jim Woodward, Bill LaPrade, and Bob Supnik.

REFERENCES

[Bhandarkar 82] Dileep Bhandarkar, "Architecture Management for Ensuring Software Compatibility in the VAX Family of Computers," *IEEE Computer*, pp. 87-93, February, 1982.

[Calcagni 84] R. Calcagni and W. Sherwood, "Patchable Control Store for Reduced Microcode Risk in a VLSI VAX Microcomputer," *Proceedings of the MICRO-17 Conference*, New Orleans, LA, October, 1984.

[Gries 84] Robert Gries and J. A. Woodward, "Software Tools Used in the Development of a VLSI VAX Microcomputer," *Proceedings of the MICRO-17 Conference*, New Orleans, LA, October, 1984.

[Kearney 84] M. Kearney, "DECSIM: A Multi-Level Simulation System for Digital Design," *Proceedings of ICCAD*, New York, October, 1984.

[Johnson 84a] W.N. Johnson, "A VLSI VAX Chip Set," *ISSCC Digest of Technical Papers*, pp. 174-175, 1984.

[Johnson 84b] W.N. Johnson, "VLSI VAX Microcomputer," *Digest of Papers*, COMPCON 84, pp. 242-246, Feb. 1984

[Levy 80] H.M. Levy and R.M. Eckhouse, *Computer Programming and Architecture: The VAX-11*, Digital Press, Bedford, Mass., 1980.

[Sherwood 84] W. Sherwood, "A Prototype Engineering Tester for Microcode/Hardware Debug," *Proceedings of the MICRO-17 Conference*, New Orleans, LA, October, 1984.

[Sherwood 81] W. Sherwood, "An Interactive Simulation Debugging Interface," *Computer Hardware Description Languages and Their Applications*, Breuer and Hartenstein, editors, North-Holland Publishing Co., Amsterdam, pp. 137-144, 1981.

[VAX 81a] *VAX-11 Architecture Reference Manual*, Revision 6.1, Digital Equipment Corp., Maynard, Mass., 1981.

[VAX 81b] *VAX-11 Symbolic Debugger Reference Manual*, Digital Equipment Corp., Maynard, Mass., 1981.

EXTERNAL AND INTERNAL ARCHITECTURE
OF THE P32, A 32 BIT MICROPROCESSOR

A. DE GLORIA

Department of Electronic Engineering
University of Genova,
16145 - Genova, Italy.

1. Introduction

This paper describes the P32, a 32-bit
microprocessor, presently under development at the
Department of Electronic Engineering of the University
of Genoa, Italy.

The P32 represents a pilot project which has two
main purposes: to study in depth the VLSI design
process; to design the core hardware for the
development of processors for in-house applications.

The development of a complex unit, such as a
microprocessor, allows to experiment several methods to
approach VLSI design and to estabilish which design
strategies are convenient to manage a VLSI project, and
which CAD tools are best suited for the design of VLSI
custom chips.

The project will provide a set of hardware
primitives to be utilized in future designs. Moreover
the processor is a way to study the internal
architecture of VLSI computers and to examine the
relation between external and internal architecture.

The P32 is characterized by orthogonality and
composability of data types, instructions and
addressing modes; these features, in conjunction with
regularity in the implementation of the instructions,
allow an easy implementation of compilers and a more

efficient code generation. The P32 provides a large uniform address space (4 giga bytes), a virtual memory capability and a basic support of Operating Systems (OS) functions. Four privilege levels, privileged instructions and support of memory access privileges provide the basic mechanisms to guarantee the security of computer system. The instruction set supports general purpose instructions and may be extended by a coprocessor capability.

The microarchitecture was designed to be expansible, to allow the updating of the processor without redesigning the entire chip, and to be modular to allow the use of some modules in the implementation of future, dedicated processors.

2. P32 External Architecture

2.1. Programming Model

The processor supports four data types: bit, byte, halfword (16 bit) and word (32 bit); either halfword or word may start at an arbitrary byte boundary and, similar to byte, they may be interpreted as either unsigned or signed data in arithmetic and logic instructions. As shown in Fig. 2.1 the programming model of the processor provides sixteen 32-bit general purpose registers. They are byte-, halfword-, and word-addressable and they may store either data or addresses. R15 is the program counter while R14 is the stack pointer. The processor provides also eight 32-bit privileged registers that can be written only in the most privileged level. These registers are used to support the management of the machine and the Operating System functions. There is also a 32-bit processor status word, subdivided into two fields, each of them 16-bit long. The most significant 16 bits can be referenced (read or written) only in the most privileged level.

The memory is organized as a linear array of four giga bytes; it is byte-, halfword- and word-addressable on even and odd address.

The instructions, which have a variable length and byte-aligned format, are composed of three fields: opcode, first operand specifier, second operand specifier, and were designed with future expansions in mind. In fact, none of the three fields is fully utilized.

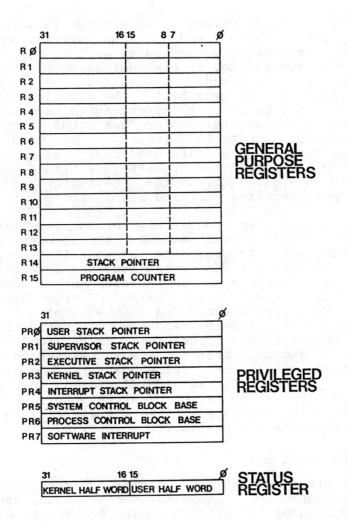

Figure 2.1 - P32 - Programming Model

The opcode may be one or more bytes long, and a special opcode extends the field to the next byte, though only one byte opcode is implemented.

The first and the second operand specifier are equal in format. The format is composed of two fields: the addressing mode field and the displacement field.

The first field may be one or more bytes long (actually only one byte is used), and contains information about the mode to address the operand as well as the register involved in addressing. In the addressing mode subfield a code is reserved to extend the addressing specification to the next byte . The displacement field may or may not be present, depending on the addressing mode, and it may be one, two or four bytes long. The microprocessor supports twelve addressing modes: register direct, register deferred, autoincrement, autodecrement, autoincrement deferred, byte-, halfword- and word displacement, byte-, halfword- and word-displacement deferred and literal.

Ortogonality and regularity have been the most important rules followed in the design of the instruction set. According to [1] each instruction can be combined with any addressing mode and any data type, in order to allow a more easy development of compilers. This choice, however, requests a large amount of hardware to be implemented and so the instruction set was necessarily limited. Fourty-four instructions are provided by the microprocessor. They are shown in Appendix A. The instruction set was designed to support only general operations, giving however some capacity to manage procedure linkage (ENTER, EXIT), to implement semaphores (bit manipulation instructions), to speed some OS basic functions (privileged instructions), and finally to guarantee protection levels.

2.2. Operating Modes

The processor has three operating modes: normal, interrupt and halt. The normal mode is associated with the execution of a process. In order to allow a structured implementation of the Operating System this mode has four privilege levels: user, supervisor, executive and kernel. The current privilege level and the previous one are specified by the CPL and PPL fields in the processor status word. There is one privileged register associated with each level, which is used as a stack pointer. Whenever the processor changes from a one level to another, R14 is updated consequently. Each privilege level has associated a certain set of the processor resources. The kernel level, the most privileged, has access to all resources.

The interrupt mode is associated with the interrupt management. It has access to all resources and also has a privilege register (the interrupt stack pointer). This mode is indicated by the IM field of the PSW.

The Halt mode is associated with a catastrophic event that inhibits the normal hardware functions.

2.3. Interrupt management

The processor has 31 interrupt priority levels subdivided in 15 software and 16 hardware. The hardware interrupts may be vectored or autovectored. A privileged register (the software interrupt register, SIR) is associated with the software interrupt, each bit of the SIR corresponds to a priority level. The software interrupts are requested by writing a "one" in the bit of the SIR corresponding to the desired level. The system control base register retains the address of the portion of memory containing the interrupt and exception vectors. There are 128 vectors.

The field IPL of the processor status word specifies the level of priority of the process running on the machine. At the end of each instruction the processor compares the IPL to the level present on the pin of interrupt request. If it is less then this one an interrupt is taken. After receiving an interrupt, and calculating the vector, the processor continues the execution at the address specified by the interrupt vector, saving in the stack its status at the interrupt reception.

2.4. Exceptions handling

The processor handles three types of exceptions: Trap: this is an exception condition caused by the execution of an instruction, occurring at the end of the instruction execution. Fault: this occurs during the execution of an instruction but allows the continuation of it once the fault cause is removed. (Page Fault, Bus Error, Address violation). Abort: this occurs during the execution of an instruction but does not allow the continuation of it. (Kernel Stack not valid, Interrupt stack not valid)

After the detection of an exception the processor saves the present status in the stack and executes the instruction stored in the location specified by the interrupt vector. The stack may have three types of format depending on the exception type. a) The short

format, equal to the interrupt stack format, is
associated to traps. b) The long format was created to
manage the fault exception, particularly for the
support of virtual memory. The processor uses the
continuation method to support virtual memory, that is
all the internal state of the processor is saved in the
stack so, once the page of memory that has caused the
fault is stored in the physical memory, the processor
resumes the status from the stack and may continue the
instruction execution. A special status word is
provided to support Operating System reserved
operations. c) The coprocessor format is reserved for
coprocessor faults and is discussed below.

2.5. Coprocessor Management

In order to extend the computational power of the
microprocessor a coprocessor capability is provided.
The processor may handle up to 128 coprocessors. A
special instruction is reserved for this purpose:

CUSTOM cp_id, cp_opcode, cp_operands

where:
cp_id is the coprocessor identifier,
cp_opcode the coprocessor instruction opcode,
cp_operands the instruction operands.

At the reception of the CUSTOM instruction the
coprocessor does not decode it, but enters the
cp_request_status whereupon the processor sends on the
address lines the cp_id, and, on the data lines, the
cp_opcode. If a coprocessor is not present, the
processor executes a coprocessor fault and the CUSTOM
instruction may be executed by software routines.
Otherwise if the coprocessor is present it becomes the
bus master and then decodes and executes the
instruction. At the end of the execution the
coprocessor sends to the processor the address of the
instruction following the CUSTOM.

A CUSTOM instruction specification may be
arbitrarily long, depending on the coprocessor
operation requested. During the execution of an
instruction some events may inhibit the current
execution. These events may be: page fault, address
violation, bus error and others. In this case the
coprocessor suspends the execution and communicates, by
a special status word and other information, the cause
of the suspension to the processor. This one will then

remove these causes, allowing the copprocessor to continue the execution. If, during the coprocessor operations, the processor receives an interrupt request, it asserts the coprocessor interrupt state, and the coprocessor may decide whether to suspend its operations and allow the interrupt service, or to continue them.

3. Internal Architecture

The need to have a general purpose processor that, with minimal adapting, may be used again in several projects of future processors was a driving factor in the design of the internal architecture of the P32. The microarchitecture of the processor was then conceived to be modular and expandable. The modularity allows the different modules of the processor to be independent; so, when it is necessary to redesign a module to satisfy new requirements, this does not involve the complete redesign of the processor. The expansibility ensures that, if new features must be added to the processor, the internal architecture does not have to be completely redesigned. The adaptability is conceived considering the possibility to support different instruction sets and formats and to support different bus interfaces. The expansibility is conceived to support an expansion of the instruction set.

In order to satisfy the above requirements the processor internal architecture is organized in three subprocessors Fig. 3.1: the Decoder Processor (DP) that decodes the instructions and sends them, in an internal format, to the Nucleus Processor; the Nucleus Processor (NP) that executes the instructions; the Interface Processor (IP) that realizes the interface with the external world and arbitrates the requests to the memory bus of the DP and NP. This organization allows the main steps of the execution of an instruction to be independent. It is then possible to have different formats of the same instruction by changing the decoding subprocessor; it is also possible to have different bus protocols by changing the interface processor. Moreover this organization allows to increment the speed of the instruction execution by the overlapping of the decoding, execution and memory referencing.

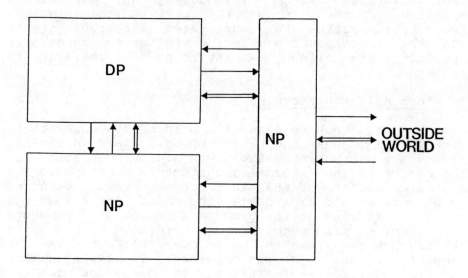

Figure 3.1 - P32 Internal Structure

3.1. Decoder Processor

The DP decodes the instructions and sends to the
NP all the necessary information to execute them. DP is
also interfaced with the IP, to which it requests to
read the instructions. The memory is always read by
words that are stored in a 64-bit buffer internal to
the DP. The DP also manages the faults caused by the
instruction reading and decoding, furnishing to the NP
the microaddress of faults-related microroutine. As
shown in Fig. 3.2 the DP is composed by four main
units:

Buffer Unit: it stores the words read by the IP. It is
byte- and halfword-addressable on even and odd
addresses, and even-addressable for word references.

Program Counters Unit: it controls the buffer status
and contains three registers: MPC is a pointer to the
word to be read by IP; BPC is a pointer to the first
undecoded byte in the buffer; PCD is a pointer to the
address of the first byte of the instruction following
the one that is being executed.

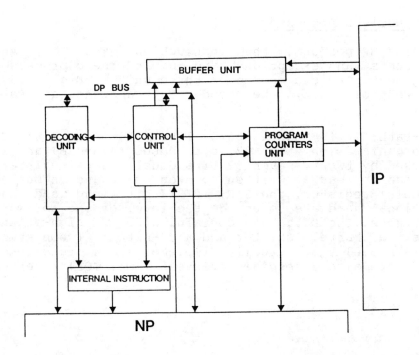

Figure 3.2 - DP internal structure.

Decoding Unit: it is composed of three tables, denoted A, B, C and of two addressing-mode-decoding subunits. Table A extracts from the opcode the information about the Alu code the number and the type of the operands, and calculates the micro-address of the micro-routines dedicated to the execution of the zero operands instructions; table B receives inputs from table A and from the first operand addressing mode decoding subunit and calculates the address of the micro-routine dedicated to the execution of the one operand instructions; table C receives inputs from table A and from the first and second operand addressing mode decoding subunits and calculates the address of the micro-routines dedicated to the execution of the two operands instructions.

Control Unit: it is a FSM dedicated to the control of the DP operations.

3.2. Nucleus Processor

The NP performs the instruction execution, the interrupts and exceptions handling and the coprocessor management. It is composed of three units (Fig. 3.3): the Control Unit, the Decoder Unit and the Data Path Unit.

Data-Path: It is Mead & Conway like [2] (Fig. 3.4). It is organized on two 32 bit buses (A and B) and is composed by twenty-seven 32 bits double-ports registers that can be read or written both on bus A and on bus B. A special register, the SIR, is 17-bit long and has additional hardware to encode the level of the software interrupts. The Data-Path contains also 2 sign-extend units, a barrel shifter and a 32-bit Alu (completely programmable) that actually implements 24 functions. There are also 2 special registers (IO1, IO2) reserved

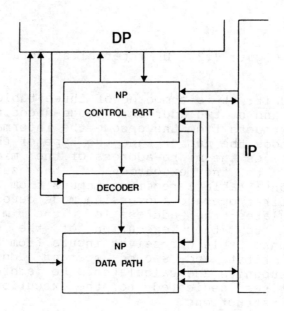

Figure 3.3 - NP internal structure.

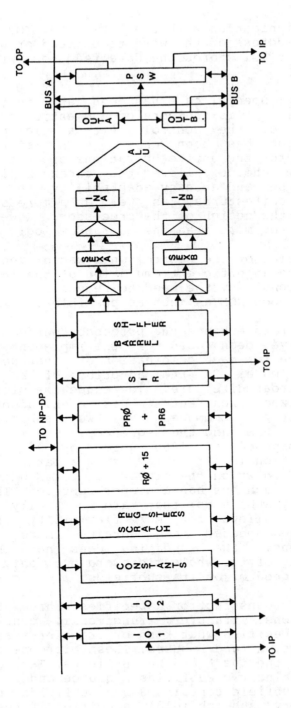

Figure 3.4 - NP Data-Path.

for the communication with the IP; the IO2 register
stores the address of the word to be read or written in
memory, while IO1 stores the data read or written by
the IP.

Control Unit: Special care has been taken in the design
of this unit. It is the core of the entire system. In
the design of the control unit a microprogrammed
implementation has been chosen. The microcode is
responsible for the entire operations of the NP. The
microprogram choice allows the expansibility of the
instruction set and an easy adaptability to different
external architectures, as well as a more structured
approach to the design of the processor. However this
choice requires more area then a random logic approach.

The design of the Control Unit was driven by the
need to have an efficient emulation of the instruction
set, and secondly by the need to reduce the dimensions
of the microcode ROM as much as possible .

In order to enhance the processor performance two
criteria have been used: 1) the implementation of a
micro-routine for each combination of instruction and
addressing mode; 2) the adoption of a one-level
pipeline microarchitecture. To reduce the microprogram
a parametrized architecture has been adopted. The
control unit is composed by two subunits: the
microprogram ROM and the sequencer. The microprogram
memory is composed of 1800 words by 75 bits. The
microinstruction is structured using a partial encoding
format type, in which the microword is subdivided in
fields where each one has a proper function. The fields
are formed by subfields that are partially encoded.
This choice allows for more flexibility; if a field
needs to be updated to contains more function
specifications, the updating does not involve a
restructuring of the whole microword but only of the
field questioned or of its subfields.

The microinstruction specifies three types of
microoperations: transfer, functional, control. The
first type specifies what kind of transfer has to be
made on the A and B buses. These microoperation are
specified by the bus A and bus B fields. Each of these
fields contains two subfields - source and destination
- and each subfield contains 3 sub-subfields: the first
(the validity sub-subfield) specifies if the operand
specification comes from the second and third subfield
or by the decoder; the second and the third subfield

specify the address of the operand and its type.

The functional microoperations are responsible to control the Alu operations. They are specified in the same way as for the transfer microoperations. Also in this case a subfield indicates if the Alu operation specification comes from the second subfield or the decoder. The control microoperations control the flux of the microprogram, and determine the address of the next microinstruction. There are sixteen control microoperations that allow a structured approach to the microprogram implementation. They allow the use of microsubroutines, conditioned jumps and of for-endfor, while-endwhile, do-until structures. The interface with memory is completely transparent to the microcode. The memory requests are sent to the IP that performs all the needed operations to access memory. The NP waits for a response of the IP and does not known the number of accesses and cycle times needed to make a reference in memory. This allows complete flexibility to adapt the NP to different memory interfaces.

Sequencer: it is responsible to execute the control microoperation. It receives inputs from the microinstruction, the data-path status word, the DP and IP status words and calculates the address of the next microinstruction. The sequencer also saves the the NP internal status at the reception of a Fault.

3.3. Interface Processor

The IP realizes the interface with the outside world and arbitrates the memory requests to the memory bus. The IP communicates with the DP through a 32 bit bus and two control lines.

The program counter unit of the DP detects when the buffer is not full and sends a request to the IP. This one if not occupied to service NP requests, reads, on the IP-DP bus the address of the word to read and makes a reference in memory. If the transaction has succeeded the IP stores in the DP buffer the word read, otherwise it communicates the failure cause to the DP.

The IP communicates with the NP through a 32 bit bus and several control lines. When the NP requests a memory reference to the IP, this one reads through the IP-NP bus the address of the word to reference in the IO2 register of the NP Data Path and the size of the data on other dedicated lines. Then, depending on the

size of the data to reference and its address, the IP
must make one or two memory accesses and adjusts
opportunely the data (read or to write). If for
example, the NP requests to read a word at an odd
address, say (n+3), where n is a multiple of 4, the IP
must take the following steps: it reads the word at the
address n, rotates the data by 8 positions to the left,
extracts the least significant byte and stores it in
the least significant byte of the register IO1 of the
NP Data-Path. Then the IP reads the word at the
address (n+4), rotates it by eight positions to the
left, extracts the three most significant bytes and
stores them in the most significant bytes of the
registers IO1. Finally the IP sends to the NP the
read-success flag. If, during an access, the external
memory controller unit sends to the processor a bus
retry, the IP restarts the memory operation. If a page
fault, a bus error or an address violation occurs, the
IP stops the memory operation and communicates to the
NP the failure.

The IP is also responsible to detect the interrupt
requests (either software or hardware), to compare
their levels with the IPL of the processor status word
and to send an interrupt request to the NP. In the
interrupt management the IP requests also the vector
number to the external unit that has requested the
interrupt. If the interrupt is autovectored the IP
derives the appropriate vector number from an internal
ROM .

The IP detects also the external bus requests and
releases the bus control to external units when neither
DP nor NP have pending memory requests.

4. Design Considerations

The project of the processor cannot be really
divided into distinct steps, but all the design phases
has been considered and carried out concurrently. Only
the external architecture design step may be separated
from the other steps but not completely. When the
microarchitecture design was started, the behavioral
description of the processor was defined only up to
70%. The microarchitecture, the logic and, in part,
the circuit designs have been performed at the same
time to take an engineering tradeoff between chip
performance and technology constraints. The final goal
was to minimize the chip area without drastically
reducing the processor performance. All the design

steps were managed using a chip planning methodology: every design choice was verified taking into account its implication at the layout level. In fact certain choices that were good from a functional point of view were rejected because they did not match with chip-planning indications. The design was carried out in a hierarchical way; the chip was fist decomposed in three modules: IP, DP, NP, and a floor-plan (Fig. 4.1) was made to evaluate the total area and to drive the design of each module. Then each module was decomposed until primitive (not decomposable) modules were reached. The modules were designed individually and connected using interconnection modules.

The design tools used are: Karl (functional simulator), Logcap (logic simulator), Spice (circuit simulator), Defasm (meta-microassembler) [3], Kic (layout editor) [4], Plaid (PLA generator), Maskap (connectivity verifier).

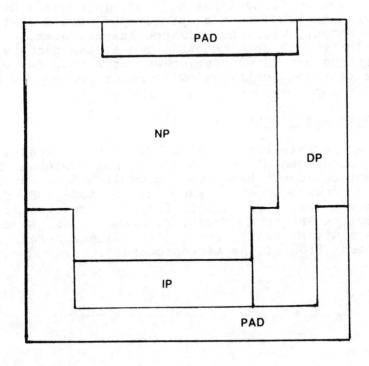

Figure 4.1 - P32 Floor-Plan

142

Karl [5] was used to verify the correctness of the
functional specification. The whole system was
simulated at the functional level in three steps: in
the first one no delays were introduced to verify the
machine functional behaviour; in the second step the
delays were considered to identify the longer paths and
to verify possible errors due to the delays, in the
third step no delay were considered and the debugging
of the microprogram was done running also some
benchmarks.

5. Conclusion

The microproccessor will be fabricated (in 1985)
using a 2.5 micron NMOS technology with two transistor
types and bulk polarization. The total area of the chip
is 6.5mm x 6.5mm, and the estimated two-phase clock
cycle length is 120 nanoseconds. The microprogram ROM
was implemented using NAND-type approach that allows
more compactness in the layout implementation. The chip
will be tested using a in-house developed
instrumentation, and a prototype board will be
implemented, with the UNIX operating system, using a C
compiler (under development). The system prototype will
be the base for the development of a computer with in-
house designed chips and will be dedicated to special
purpose applications.

6. Acknowledgement

A special thank goes to Prof. P. Antognetti for
his encouragement and suggestions; without him the
project could not have been accomplished. I would like
to acknowledge the support of SGS-ATES Co.; in
particular Dr. P. Rosini for his collaboration. A
special thank to M. Buzzo, M. Gismondi, R. Lanzarotti,
S. Pezzini, M. Valle, for their collaboration in the
implementation of the microprocessor.

Appendix A

P32 instruction Set.

Every instruction is represented by a mnemonic code that specifies what type of operation must be performed. If the instruction requests some operands, and if the operand data type is not implicit, the operand is represented by a mnemonic code. In the following, the postfix i is used to indicate one of the three types of data: byte, halfword and word. The addressing is specified by:

gend the operand is of type destination, i.e the literal addressing mode cannot be used.

gens the operand is of source type, all the addressing modes may be used

gena the operand is of address type, the literal and register direct addressing mode cannot be used.

Arithmetic and Logical Instructions

Opcode	Operand Specifications	Operation
ADD.i	gens, gend	add binary
ADDC.i	gens, gend	add binary with carry
AND.i	gens, gend	logical and
CLR.i	gend	clear
CMP.i	gens, gend	compare
CMPL.i	gens, gend	logical compare
DIVU.i	gens, gend	unsigned binary division
DIVS.i	gens, gend	signed binary division
EOR.i	gens, gend	logical eor
MULU.i	gens, gend	unsigned binary multiply
MULS.i	gens, gend	signed binary multiply
NEG.i	gend	negate
NOT.i	gend	logical not
OR.i	gens, gend	logical or
SUB.i	gens, gend	binary subtract
SUBC.i	gens, gend	binary subtract with carry
TST.i	gens	compare with zero

Data Movement Instructions

ESMOVE.i	gens, gend	extend sign and move
LDSW.w	gens	load processor status halfword
MOVE.i	gens, gend	move data
MOVEcc.i	gens, gend	move on condition
ENTER	mask, disp.i	save registers and allocate stack frame
EXIT		restore registers
STSW.h	gend	store processor status halfword
SWAP.i	gend	swap data

Bit Manipulation Instructions

BITCHG.i	gens, gend	bit test and change interlocked
BITCLR.i	gens, gend	bit test and clear interlocked
BITSET.i	gens, gend	bit test and set interlocked

Program Flow Control Instructions

BPT		breakpoint trap
CALL	gena	call subroutine
CHPLx	code number	change privilege level
DJMP	gens, gend	decrement and jump
JMPcc	gena	jump on condition
REI		return from exception or interrupt
RTS		return from subroutine

Coprocessor Management Instructions

CPREGR	cp_Id, cp_Reg, gend	cp register read
CPREGW	cp_Id, cp_Reg, gens	cp register write
CUSTOM	cp_Id, cp_opcode, cp_op	coprocessor request

Privileged Instructions

LDHPC		load process hardware context
LDPR	gens, prnumber	load privileged register
LDSW.i	gens	load processor status word
STPHC		store process hardware context
STPR	prnumber, gend	store privileged register
STSW.i	gend	store processor status word

Miscellaneous Instructions

PROBER	mode.b, gena	probe address for reading
PROBEW	mode.b, gena	probe address for writing

References

1. W. A. Wulf, "Compilers and Computer Architectures", <u>IEEE Computer</u>, (July 1981).

2. C. A. Mead and L. A. Conway, <u>Introduction to VLSI Systems</u>, Addison-Wesley Reading , Mass (1980.).

3. M. Mezzalama and P. Prinetto, "Firmware description languages for a microprogram meta-assembler", in <u>Computer Hardware Description Languages and their Applications</u>, North-Holland Publishing Company (1981).

4. K. H. Keller, "Kic: A graphic editor for integrated circuits", in <u>MS Rep.</u>, , Univ. of California, Berkeley (June 1981).

5. R. W. Hartenstein, "Karl-II eine Sprache zuz Spezifikation beim Entwurf Kunden spezifischer Digital bansteine", <u>Angenwandte Informatik</u>, (December 1982).

A HIGHLY REGULAR PERIPHERAL PROCESSOR

Wolf-Dietrich Moeller and Gerd Sandweg

Corporate Laboratories for Information Technology
Siemens AG, Muenchen, Germany

ABSTRACT

The architecture of an experimental 8/16-bit peripheral processor is described. The chip was fabricated in a 2-μm NMOS technology with two metal layers. There are about 300 000 transistors on a chip area of 105 mm^2. The average time for a read-modify-write instruction is 200 ns. A highly modular and regular design style and some automatically generated layouts resulted in a short design time.

The following subjects are discussed in more detail:
Goals of the development, specification of the processor, impacts of the chosen technology, global architecture, remarkable features of the data and control unit, test results and experiences.

1   GOALS OF THE DEVELOPMENT

The rapid growth of VLSI-technology requires new methods to cope with the complexity of the designed circuits. After our research group had successfully developed a 32-bit execution unit as a model of a VLSI chip /1/, we were looking for a real VLSI chip with more than 100 000 transistors. We found a suitable object in a peripheral processor from our Data Processing Division. This processor, called PP4, has to control the data flow between mass storage devices (e.g. discs) and a mainframe computer.

Our goal was to integrate the kernel of this processor on one VLSI chip. During the project it became evident that a stand-alone VLSI chip would not result in economic advantages comparing a solution based on a printed circuit board with standard IC's. Economic advantages could be reached only by designing a whole system in a VLSI style. But in spite of this problem we decided to go ahead with our PP4 chip claiming it to be an experimental chip with a realistic task. We used this chip as a test vehicle for advanced design methods, some new architectural concepts and a risky technology /2,3,4/.

2   SPECIFICATION OF THE PROCESSOR

The processor is intended to control the data flow between high speed peripheral devices and the CPU of a mainframe computer.

The requirements for such a processor are very challenging:

- High data rates (up to 5 Mbyte/s), resulting in a cycle time in the range of 200 ns.
- Real-time operation, this means worst-case conditions for timing must be assumed.
- Flexible programming, as the specifications of future peripheral units cannot be foreseen.
- Powerful instructions for 8- and 16-bit data.

General purpose microprocessors, even the most advanced ones, are not fast enough to fulfill these requirements, therefore a special processor with some dedicated hardware is necessary. To meet these stringent requirements, the following decisions were taken with respect to the architecture of the processor:

- Overlap of instruction fetch and execution.
- Instruction execution by a small (but variable) number of steps.
- Horizontally coded instruction format with 32 bit to speed decoding.
- Orthogonal simple structure of the instruction format (RISC architecture /5/).
- Registers and data path operating in 16-bit mode and 8-bit mode (low byte only).
- Large register set of 32 general purpose registers, 5 special registers and 192 addressable external registers.
- Memory address space of 64 Kwords.
- Part of the memory is placed on the processor chip (1.5 Kwords).
- Memory locations may be addressed directly or indirectly.
- Data (e.g. parameter tables) can be stored in memory.

3   IMPACTS OF TECHNOLOGY

The processor chip was fabricated in an advanced NMOS technology with 2-μm structures. There is a single low-ohmic polycide layer for gates and short interconnections. Most of the signal lines and the power supply lines run in the first metal layer.

New for us was the use of a second metal layer. Because of the coarser design rules for this layer it could not be used unconditionally. We gained by it a denser design only in some special cases. We used the second metal layer mainly for power supply lines (sometimes redundant with lines in the first metal layer) and for lines in parallel to polycide lines to reduce resistance and signal delay. To improve testability we provided probe points in the upper metal layer for electron beam test methods /6/. Finally the second metal layer is helpful in the redesign task.

The first layout estimations showed a die area of about 100 mm^2. Because of this large area it was evident that we could expect only a very poor yield. But since the PP4 is an experimental chip we were allowed to keep on this risky way.

Considering the supply voltage of 5 V and the number of about 300 000 transistors it is necessary to design the transistors carefully to limit the power consumption. We expect a value in the range of 3 to 4 W.

A severe constraint is the number of pins. For the PP4 we need a package with more than 152 pins. We have chosen a rather expensive square 224-pin package with a multilayer board inside. For a large production volume this is perhaps not the right choice but for an experimental chip it is tolerable.

4   GLOBAL ARCHITECTURE OF THE PP4 PROCESSOR

Main goal in the design of the PP4 processor was to achieve a very regular and modular structure. This should be accomplished by use of ROM, RAM, PLA and highly repetitive structures wherever possible. Apart from some minor units for sychronizing and interrupt, the processor is devided into three sections: data unit, control unit and memory unit (Fig. 1).

The data unit consists of a data port, an execution unit and a sequencer. The data port serves as an interface to external registers and for data to and from the memory. The execution unit handles all data manipulation. The sequencer provides the next instruction address. It is often located in the control section. But as the main task of the sequencer is 16-bit address manipu-

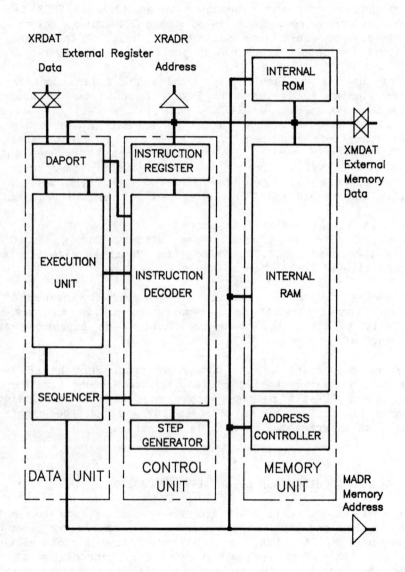

Fig. 1   PP4 block architecture

lation, its structure is very similar to the structure of the other modules in the data unit. Therefore the sequencer fits very well into the data unit.

The control unit has to provide the control signals for all other units of the processor. It is normally considered to be the most irregular part of a processor. But due to the modular instruction format and excluding the sequencer, a very regular 2-stage decoding scheme with PLA-like structures could be realized.

Address space for memory is 64 Kwords. One word consists of 32 data bits and 4 parity bits. The whole addressable memory would be too large to be placed on the processor chip itself. But to reduce long access times to the memory we tried to provide as much internal memory as possible. Thus a ROM section of 512 words for bootstrap and selftest programs avoids the hardware overhead of an external ROM with control logic. A RAM of 1024 words allows rapid access to parameter tables and to often used subroutines. The external memory is accessed via the 16-bit memory address bus and the 36-bit data bus.

Both register and memory address space contain internal and external storage locations.

5   DATA UNIT

5.1 Structure of the Data Unit

The data unit is composed in a matrix structure by bit slices and function slices. This concept has proved to be very effective when taking processing speed, area and design effort into account. The recommended architecture is called a data path /7/ and consists of several function slices that are built together without any additional wiring. This is because the cells of the function slices have an integrated bus system, in our case a 2-bus system (Fig. 2). The data lines run in the metal layer to minimize signal delays. The data lines are crossed by the control lines in the polycide layer.

Each function slice is built with bit cells arranged in a serial manner. So for each function slice only one bit has to be designed.

5.2 Busses and Timing in the Data Unit

The instruction set of the PP4 consists mainly of 1-operand and 2-operand instructions. Thus a 2-bus system (A-Bus and B-Bus)

152

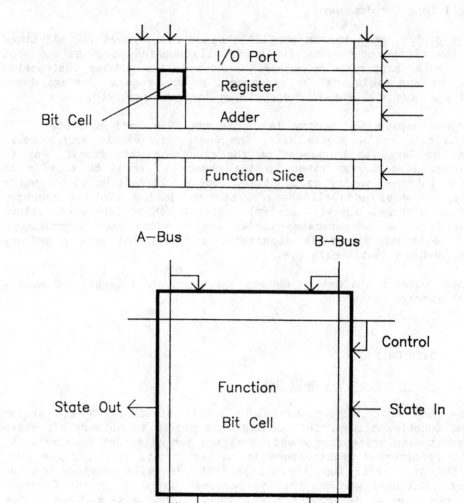

Fig. 2   Slice technique

and dual-port registers for the internal register file are adequate regarding the speed requirements. In an ordinary 2-operand instruction the two operands are fetched simultaneously from the registers via the A-bus and the B-bus. The result is written into the destination register reusing the A-bus.

There were wishes for some 3-operand instructions. But the provision of the 3rd operand via a special register was discarded because of introducing too much irregularity into the architecture. Thus 3-operand instructions have to be replaced by a sequence of 2-operand instructions.

To keep the clock rate at a moderate level, a timing scheme was selected with phases as long as possible. For an ordinary register-to-register operation two phases are necessary. This allows for switching between read and write mode. The data manipulation can be executed asynchronously during the read and write phase (Fig. 3). Input data for the manipulation function slices is held constant during phase 2 by input latches.

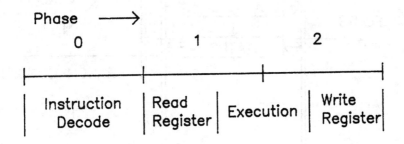

Fig. 3   Timing of register-to-register instruction

5.3 Function Slices

The data path consists of 6 function slices (Fig. 4).

At the upper end of the data path the data port (DAPORT) forms an interface to the external registers and to the memory. The data port transforms 32-bit memory words into 8- and 16-bit words, and vice versa. Additionally parity check or parity generation is performed. This introduces slight irregularities in the design, as 9 bits have to be accommodated in the width of 8 bit slices.

The register file (RFILE) consists of dual-port registers. There are 32 general purpose and 5 special registers, one of which is

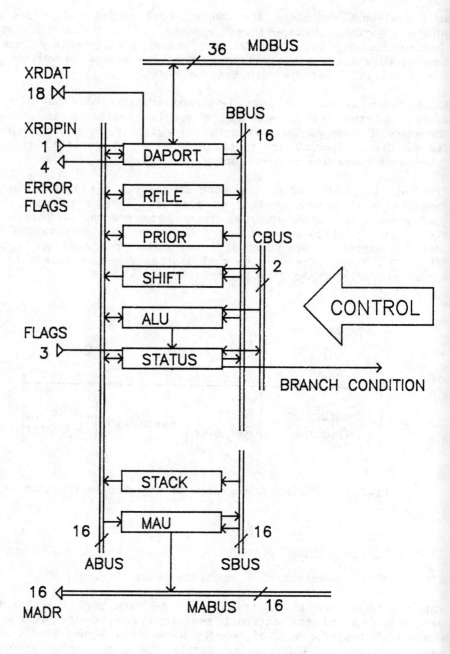

Fig. 4   Structure of the PP4 data unit

a hardwired zero. Two different registers may be read out to the A-bus and B-bus simultaneously.

The priority encoder (PRIOR) determines the most significant "1" in a 8-bit or 16-bit word and codes its position in a 5-bit number. A fast priority encoder may be designed as a PLA structure. But the problem is how to integrate a PLA into a data path structure. We solved this by merging the AND- and the OR-plane inside the data path (Fig. 5).

The AND-plane fits easily into the bit slice structure if the input lines are running in parallel to the data bus. Consequently the product term lines have to be realized in the polysilicon layer, which is not too critical since we use the low-ohmic polycide. There is enough space in the bit cells for the transistors of the OR-plane or alternatively the pullups of the product term lines. So it is possible to put the OR-plane over the AND-plane. To assemble this merged AND-OR-plane nine different basic cells with equal height must be prepared.

The barrel shifter (SHIFT) executes various multiple bit shift operations in one cycle.

The function slice is divided into an input preparation logic and into the shifter matrix (Fig. 6). The control lines running diagonally across the shifter matrix are realized in the second metal layer to permit an area saving layout with simple crossings.

In contrast to a conventional crossbar switch the shifter matrix uses 2 inputs instead of one. The desired shift functions (e.g. arithmetic shift, logic shift, left shift, right shift, rotate) are performed by setting both inputs correspondingly (e.g. cyclic rotation is done with the same data word at both inputs). These inputs are controlled by the input preparation logic.

This approach has several advantages. Shifting in "ones"," zeros" and status bits is fast (it is done parallel) and needs no additional circuits. Word- and byte-modes are implemented easily. Finally a low transistor count, high regularity and minimum wiring effort is achieved.

The arithmetic-logic unit (ALU) performs the standard logic and arithmetic functions. For symmetry both subtract functions (A-B, B-A) are provided. Increment, decrement and masking operations are possible with all powers of 2.

The time critical path in an ALU is the carry chain. Due to the chosen timing scheme the use of a precharge Manchestertype carry chain was not possible. Instead of this, a simple 2-bit carry look-ahead adder with NOR-gates showed to be most effective and

156

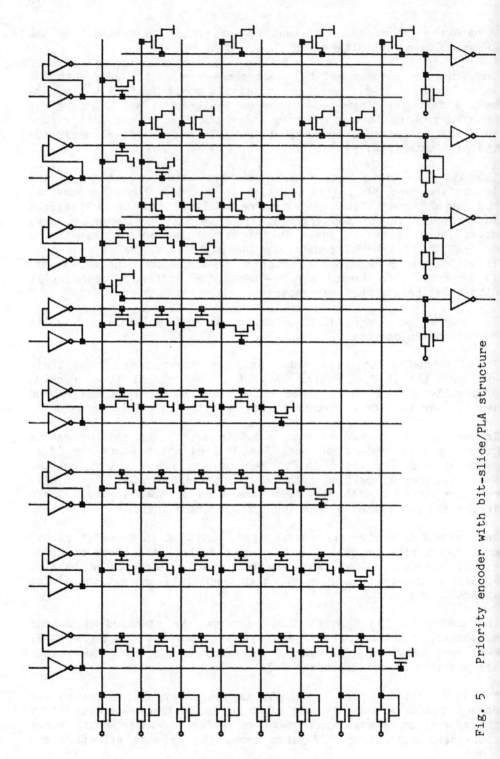

Fig. 5   Priority encoder with bit-slice/PLA structure

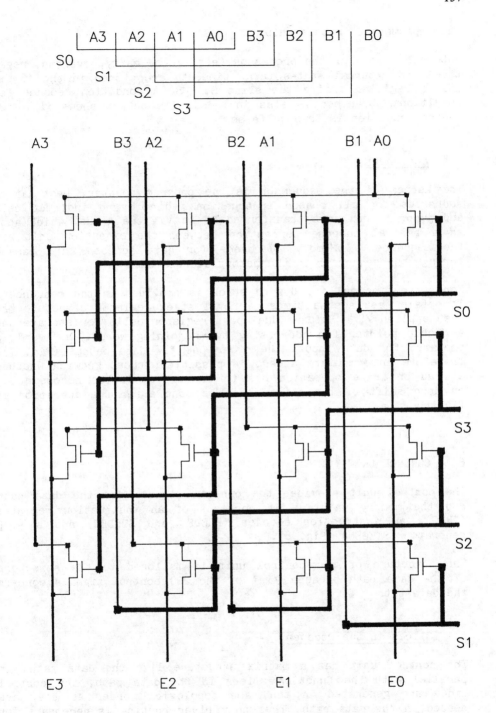

Fig. 6   Barrel shifter matrix (4 bit)

was implemented in this chip.

Internal states of the processor (e.g. zero, carry, overrun, negative) and external states (e.g. timeout) are stored in the 16-bit status register inside the slice STATUS. A condition decoder for conditional branches is also included. Its output shows if a selected condition is true or false.

5.4 Sequencer

The kernel of the sequencer is the program counter. Destination addresses for all branch instructions (direct and indirect call and jump, return) are transferred to it via the A-bus, which acts as a general purpose connection between the execution unit and the sequencer. Memory addresses for data accesses are passed through the sequencer to the memory address bus.

For subroutine calls a 8-word stack is provided in the sequencer. To obtain fast access times, a novel stack register structure has been applied /8/. It uses dual-port register cells for storage and a double pointer for addressing two adjoining words. One word is prepared for data input, the other one for data output (Fig. 7). Therefore an immediate access is possible without knowing whether a read or write operation has to be performed. The necessary up or down shift of the pointer can be done after the operation is finished.

6 CONTROL UNIT

The control unit provides the control signals for the data path and the memory unit. It consists of an instruction register (IREG), an instruction decoder (PREDEC and STEDEC) and a step generator (STEGEN) (Fig. 8).

The structure of the control unit is tailored to the structure of the data path because 95 % of the 200 control signals concern the data path.

6.1 Slices in the Decoder

The control unit has a matrix structure like the data path. In parallel with the function slices in the data path, the control lines are generated in the same topological order as they are needed in the data path. Thus only river routing is necessary for interconnection. Each control line has its decoding and timing logic in an adjoining region of the decoder. Single function sli-

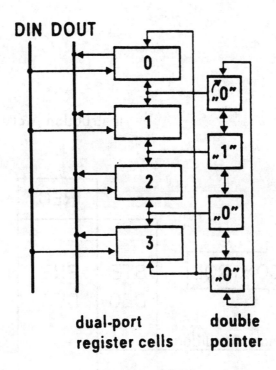

dual-port
register cells

double
pointer

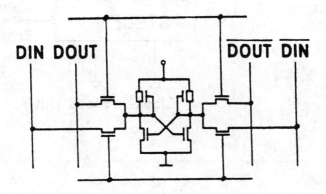

Fig. 7   Stack with dual-port register cells and double pointer

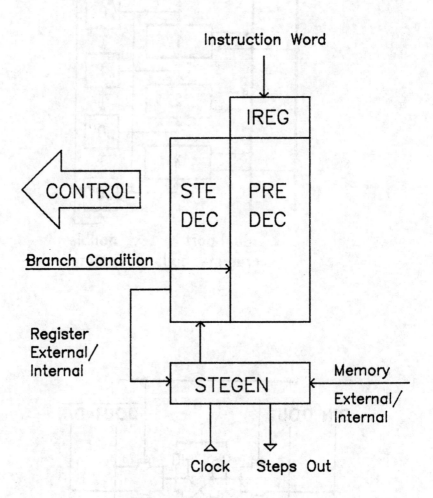

Fig. 8   Structure of the PP4 control unit

ces may be added or omitted with local influences only. Even geo-
metrical fitting between data path and decoder can be achieved
easily by adding dummy slices in the data path or in the decoder,
depending on which part must be stretched.

6.2 Instruction Decoding

At the beginning of each instruction the actual instruction code
is transferred to the output of the instruction register. From
there the control signals for the data path and the memory are
obtained by a 2-stage decoding scheme. The first stage is a predecoder (PREDEC). The output of the predecoder is stable from
the end of the first step to the end of the whole instruction.
The final timing of the control lines is accomplished in the second stage, the step decoder (STEDEC). Each output of the predecoder is gated with the appropriate step.

Design studies showed that the best predecoder structure with
respect to area and flexibility is a large PLA with a single OR-
line instead of a complete OR-plane /9/. The input lines of the
PLA are made of polycide. They are covered and connected with
parallel lines in the second metal layer to reduce the RC-delay
times on these long input lines (5.5 mm !). Product term lines
run in the first metal layer. About 500 product terms were needed
to decode the instructions. Only less than 5 % of all product
terms could have been used a second time. So it is better to spend
the area for these "redundant" product terms and to reduce the
OR-plane to a single OR-line instead of increasing area by using
an OR-plane. The AND-plane and the OR-line are realized by NOR-
Gates.

The task of the step decoder is to gate the outputs of the predecoder with the step signals and to build sum terms to form the
final control signals. Again we have an AND-OR-structure. But
because the step decoder has only few gate functions it can be
realized in a simpler way just using the first metal layer for
the step signal lines and polycide for the product term lines.
The OR-plane is reduced to an OR-line again (Fig. 9).

The instruction decoder was generated fully automatically from
an alphanumeric input file, which contains the names and the topological order of the control lines together with the decoding
and timing information. The PLA parts were generated with modified
PLA generators, while placing, spacing and wiring was done by
an automatic router.

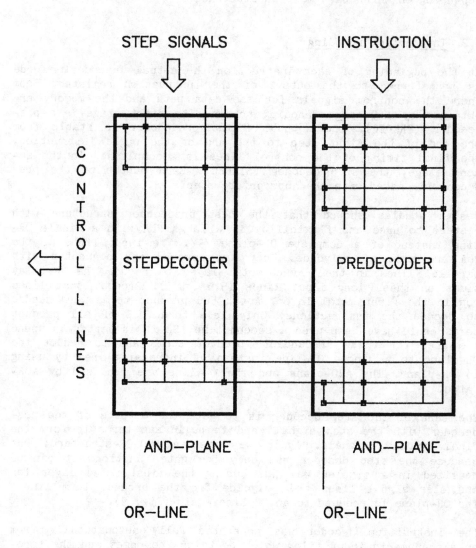

Fig. 9   Instruction decoding with PLA structures

6.3 Step Generation

To exploit the short access time to internal storage, a variable-length timing scheme was introduced (timing steps). Depending on the type of instruction several steps of the full cycle are omitted. The maximum number of steps per instruction cycle is 9 for an instruction with 2 accesses to the external memory. The minimum of 3 steps occurs in an internal register-to-register operation with next instruction in the internal memory. The time for one step is 40 ns, thus one instruction takes between 120 and 360 ns (Fig. 10). The average instruction time is 200 ns.

All steps for the processor are delivered from the central step generator (STEGEN). It is basically a finite state machine with a state register and 3 PLA's for control and generation of non-overlapping steps. It is clocked by an external master clock with a period of 40 ns. The processor may be halted at any time by stopping the clock or at the end of any instruction by a gating signal.

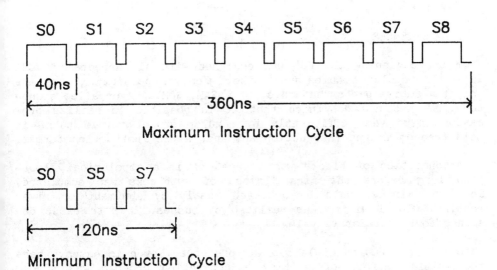

Fig. 10   Longest and shortest instruction cycle of the PP4

7   MEMORY UNIT

The addressable memory space is 64 Kwords. Each word consists of 32 data bits and 4 parity bits.

The internal ROM contains 18 Kbit, organized in 512 words by 36 bit. This storage capacity is enough for selftest and bootstrapping.

The internal RAM contains 36 Kbit, organized in 1024 words by 36 bit. To obtain fast access time and compability with the technology for logic circuitry a static 6-transistor cell was chosen for the RAM /10/.

A memory address controller monitors the addresses on the memory address bus and enables the appropriate sections of internal and external memory.

The total chip size of the PP4 is about 105 mm^2. Two thirds of this area are occupied by RAM and ROM. As yield problems were expected for this experimental chip, the ROM and/or RAM sections may be switched off by test signals to allow the operation of the PP4 with external memory only. Additionally the RAM and ROM may be tested separately by direct access via the memory address bus and the data bus.

8   EXPERIENCES

From the beginning the PP4 was designed with all aspects of VLSI in mind. Regular modules were taken whenever possible. There are only few active parts that have not a RAM, ROM, PLA or slice structure. We are now convinced that it is possible to realize processor chips very efficiently by regular structures if there is some freedom in the instruction set and in the instruction format.

Interconnection of blocks was thought of in a topological floorplan long before the first layout of subblocks was completed. So local wiring could be achieved mostly by abutment or river routing. Global wiring was realized by busses. In both cases design effort is relatively low.

Automatic generation of layout was preferred, even in cases, where the initial effort of writing a generation program was higher than the drawing of the same layout on a graphics workstation by hand. Register-transfer and gate-level simulation were employed at all stages of the design. Errors found by simulation could be corrected easily due to the generation programs and the regular and hierarchical structure of the design. Optimization cycles, in particular in the control unit, were possible only because of the use of generation programs. Optimizing the layout by hand would have involved too many steps and would have created too many new errors to be acceptable.

We got first silicon in Oct. 83. A chip microphotograph is shown in Fig. 11. The tests indicated design errors in some central parts (e.g. step generator, RAM decoder). Most of these errors would have been found by an electrical rule check program. But at the time of finishing the layout the available check programs could not handle the huge amount of data. We have got better programs now, but there is still the problem of too long computing times. Nowadays most of the CAD tools do not exploit the hierarchy going along with modern architectures. Taking advantage of this hierarchy to reduce data amount and computing time is a must for future work.

To estimate the real design effort for the PP4 is rather difficult because in our research group there were other activities not related directly to the PP4 project. Over a period of two years about 5 engineers were involved in specification, architecture, circuit design and layout of the PP4. One engineer has tested the chip and corrected all errors found within half a year. We expect silicon of a redesigned version in the fall of 1984.

166

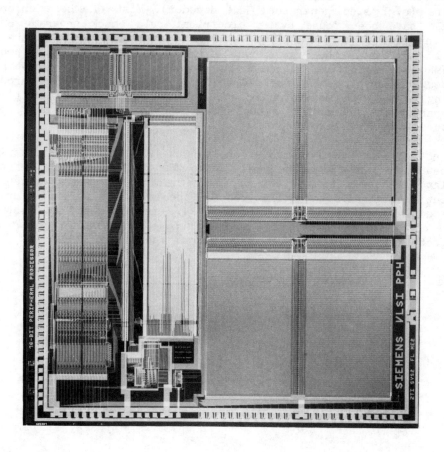

Fig. 11   Microphotograph of the PP4

REFERENCES

/1/ Pomper, M.; Beifuss, W.; Horninger, K.; Kaschke, W.: A 32-bit Execution Unit in an Advanced NMOS Technology. IEEE Journal of Solid-State Circuits, Vol. SC-17, No. 3, June 1982.

/2/ Pomper, M.; Augspurger, U.; Mueller, B.; Stockinger, J.; Schwabe, U.: A 300 K Transistor NMOS Peripheral Processor. EESCIRC 83, pp. 73 - 76, Lausanne, Switzerland, Sept. 1983.

/3/ Pomper, M.; Stockinger, J.; Augspurger, U.; Mueller, B.; Horninger, K.: A 300K Transistor NMOS Peripheral Processor. Paper to be published in the IEEE Journal of Solid-State Circuits, vol. SC-19, no. 3, June 1984.

/4/ Moeller, W.D.; Sandweg, G.: The Peripheral Processor PP4 - A Highly Regular VLSI Processor. Proceedings of the 11th Annual International Symposium on Computer Architecture, pp. 312 - 318, Ann Arbor, Michigan, June 1984.

/5/ Patterson, D.A.; Sequin, C.H.: The RISC II Micro-Architecture. Proceedings IFIP TC 10/WG 10.5 Int. Conf. on VLSI, pp. 349 - 360, Amsterdam, New York, Oxford, North Holland, 1983.

/6/ Fazekas, P.; Feuerbaum, H.-P.; Wolfgang, E.: Scanning Electron Beam Probes VLSI Chips. Electronics, July 14th 1981, pp. 105 - 112.

/7/ Mead, C.; Conway, L.: Introduction to VLSI Systems. Reading, MA, Addison-Wesley, 1980.

/8/ Schallenberger, B.: Stack mit sofortigem Schreib- und Lesezugriff. Patent pending.

/9/ Stockinger, J.; Wallstab, S.: A Regular Control Unit for VLSI Microprocessors. Proceedings ESSCIRC 83, pp. 183 - 186, Lausanne, Switzerland, Sept. 1983.

/10/ Soutchek, E.; Horninger, K.: Wortweise organisierter 36-Kbit statischer Speicher für VLSI-Prozessoren. Proceedings NTG-Fachtagung Großintegration, pp. 52 - 54, Baden-Baden, West-Germany, March 1983.

This work was supported by the Technological Program of the Federal Department of Research and Technology of the Federal Republic of Germany (Project Number NT 2588/6). The authors alone are responsible for the contents. The authors are with the Corporate Laboratories for Information Technology, Siemens AG, Muenchen, Germany.

PART II:
VLSI ARCHITECTURES FOR IMAGE PROCESSING

SCAPE: A VLSI CHIP ARCHITECTURE FOR IMAGE PROCESSING

R.M. Lea

Brunel University
Uxbridge. England

1. INTRODUCTION

The Computer Architecture group at Brunel University is engaged in research leading to the specification and design of low-cost microelectronic SIMD (Single-Instruction control of Multiple-Data streams) parallel-processing modules for the cost-effective execution of a wide range of real-time "image processing" algorithms.

The research project is dominated by the specification, design and fabrication of the SCAPE (Single-Chip Array Processing Element) chip, a 150K transistor CMOS chip which is fully optimised for the cost-effective implementation of such algorithms [1,2].

1.1. Image processing algorithms

The term "image processing" is widely used and there is some confusion as to its precise meaning. Indeed, the term is often used to cover the first two and all three of the following classes of image-related processing

(1) image-independent pre-processing: including algorithms for "image restoration", to compensate for the known imperfections of the image source (eg. TV camera), and all forms of "image enhancement" from smoothing (eg. for noise reduction) through sharpening (eg. for edge detection) and statistical analysis of grey-level detail (eg. for histogramming, contrast stretching and thresholding)

(2) image-dependent processing: from "image segmentation" (to partition the image into regions of interest or to discriminate meaningful shapes) through segment labelling and shape manipulation (eg. erosion, dilation and thinning of binary images) to feature extraction and analysis

(3) image understanding: from shape classification (eg. structure matching) through shape description to the final interpretation of image content, covering all aspects of "pattern recognition".

In general, the above classes (1) and (2) involve spatial and temporal processing of two-dimensional arrays of luminance values, whereas class (3) involves the relational processing of linked-lists of point-vectors. In fact, this processing difference is so profound, that "image processing" (and especially "image analysis") often refers only to the algorithms in classes (1) and (2) and "image understanding" is normally regarded as a totally different field.

Algorithms in classes (1) and (2) involve "global-to-point", "window-to-point" and "point-to-point" image transformations, in which pixel "points" can be represented as an n-bit grey level or a 1-bit binary level. "Global" and "point" operations can be regarded as an extension of "window" operations. Consequently, a fundamental requirement of "image processing modules", for the execution of algorithms in classes (1) and (2), is the cost-effective support of "local window" operations.

1.2. Image processing modules

An "image processing module" is an intelligent peripheral device, dedicated to the high-speed execution of image processing algorithms under the control of a "host" computer. Typical image processing modules comprise four major functional blocks, these being the

(1) Image Store (IS), providing buffer storage for a sequence of video M_i x N_i pixel "frames"

(2) Patch Processing Module (PPM), supporting high-speed (viz. parallel) processing of a M_p x N_p pixel "patch" transferred from a selected "frame"

(3) Micro-controller (MC), supporting "patch" selection and transfer and low-speed (viz. scalar) processing of intermediate results

(4) Input-Output Interface (IOI) supporting independent digitised "video-input" (eg. from a TV camera), "video output" (eg. to a

display monitor) and a communications link to the host computer.

In operation, the "host" requests the execution of one or more processes (viz. selected algorithms or sets of algorithms) on an Ms x Ns pixel "sub-image" of a selected "frame". Such "sub-image processes" are executed in one or more sequences of "patch processes".

"Patching strategies" often include sequencing Mp x Np or Ms x MpNp/Ms or MpNp/Ns x Ns pixel "patches" over the "sub-image", according to the selected algorithm.

1.3. Real-time image processing

For real-time image processing, all "sub-image processing" must be completed within a single "frame-time" (viz. 40ms for interlaced-raster TV and 50Hz mains frequency). Consequently, the

$$
\begin{array}{l}
\text{Maximum "patch processing time"} \\
\text{(including load and dump)}
\end{array}
= \frac{40MpNp}{MsNs} \text{ ms} \quad\text{---------------- (1)}
$$

Consider, as an example, the execution of "noise reduction" (eg. median filtering algorithm), "edge sharpening" (eg. high-pass filtering (viz. convolution) algorithm) and "contrast enhancement" (eg. histogram equalisation algorithm) within a single "frame-time".

Assuming a 512 x 512 "sub-image" (Ms x Ns), a 64 x 64 "patch" (Mp x Np) and a similar "patch processing time" for each algorithm, the maximum "patch processing time", for this simple sequence of only three "sub-image processes", is 0.156ms, which is equivalent to only 38ns per pixel. Clearly, a very high-speed "patch processor" is required.

1.4. Spatial convolution

The convolution of a K x K kernel (or "weight map") with an input image X to produce an output image Y is given by the following expression

$$
Y_{i,j} = \sum_{m=i-k}^{i+k} \sum_{n=j-k}^{j+k} w_{m,n} X_{i-m,j-n}
$$

where $k = \dfrac{K-1}{2}$ and K is odd

and $X_{i,j}$ is the luminance value of the input image location (i,j)
$w_{m,n}$ is the weight value of the kernel location (m,n)
$Y_{i,j}$ is the luminance value of the output image location (i,j).

1.5. Spatial convolution: SISD implementation

Assuming an Mp x Np patch with the origin (Oi,Oj) and an SISD (Single-Instruction control of a Single-Data stream) implementation of K x K "local window" processor, spatial convolution can achieved with a sequential algorithm of the form

```
for i := Oi to Oi+Mp-1 do
  for j := Oj to Oj+Np-1 do
    begin
      s := 0;
      for m := -k to k do
        for n := -k to k do
          s := s + w[m,n] * X[i-m,j-n];
      Y[i,j] := s
    end
```

where s is the sum held in the accumulator.

The performance of this SISD implementation of spatial convolution can be summarised as follows

$$\text{Time for K x K convolution over a Mp x Np "patch"} = \frac{MpNpK^2}{\emptyset} \qquad (2)$$

where the clock-rate $\emptyset$ is derived from "register load" (T_l), "multiply" (T_m) and "add" (T_a) delays as follows

$$\emptyset = \frac{1}{T_l + T_m + T_a} \qquad (3)$$

Assuming a 10MHz clock-rate, for a high-speed "state-of-the-art" SISD processor, Table 1 evaluates equation (2) for common "window" sizes.

K x K	64 x 64 "patch"	512 x 512 "sub-image"
3 x 3	3.7ms	236ms
5 x 5	10.2ms	655ms
7 x 7	20.1ms	1285ms

Table 1. Spatial convolution times for an SISD processor with $\emptyset$ = 10MHz

Significantly, equation (2) does not allow for "patch" load and dump operations. Hence, load/dump times provide an interesting reference for the data of Table 1. Assuming a 10MHz pixel-rate, the corresponding "patch" and "sub-image" load/dump times are 0.4ms and 26.2ms respectively.

Clearly, SISD processors are not capable of achieving the 40ms target of real-time image processing for, even a simple, spatial convolution. Indeed, when more than one algorithm must be executed within this time constraint, the SISD processor is woefully inadequate.

1.6. Spatial convolution: bit-serial SIMD implementation

Attempts, to achieve higher performance for more flexible real-time image processing, have led to many proposals based on bit-serial SIMD (Single-Instruction control of Multiple-Data streams) processors for this application. In fact, this field of parallel processing computer architecture is already well advanced, with major contributions including Unger's machine [3], Westinghouse's Solomon computer [4], UCL's CLIP [5], ICL's DAP [6] and Goodyear's MPP [7]. A significant feature of such proposals is the development of special-purpose parallel processing LSI chips; for example, custom-designed 8-pixel nMOS chips for the CLIP and MPP projects and a 16-pixel gate-array for the DAP project. More recently, NTT's AAP chip [8] and GEC's 32-pixel GRID chip [9] are exploring the potential of VLSI for this application.

Assuming a typical bit-serial SIMD implementation of a K x K "local window" processor and an Mp x Np patch with the origin (Oi,Oj), spatial convolution can achieved with a parallel processing algorithm of the form

```
forall i,j do Y[i,j] := 0;
for m := -k to k do
  for n := -k to k do
    forall i,j do Y[i,j] := Y[i,j] + w[m,n] * X[i-m,j-n]
```

where the "multiply-add" is performed bit-serially as follows

```
for b := 0 to p-1 do
  if w[m,n,b] = 1
    then for l := 0 to q-1 do
      Y[i,j,b+l] := Y[i,j,b+l] + X[i-m,j-n,l]
```

for p-bit pixels and q-bit weights.

The performance of this bit-serial SIMD implementation of spatial convolution can be summarised as follows

Time for K x K convolution
over a Mp x Np "patch" $= \dfrac{pqK^2}{\emptyset} + T_s$ ---------------- (4)

where the clock-rate $\emptyset$ is defined as follows

$$\emptyset = \frac{1}{2T_r + T_a + T_w}$$ -------------------------------------- (5)

and Ts represents shifting delays for non-neighbour communication.

Assuming 8-bit pixels and weights and a 10MHz clock-rate, for a typical bit-serial SIMD processor, Table 2 evaluates equation (4) for common "window" sizes.

K x K	64 x 64 "patch"	512 x 512 "sub-image"
3 x 3	230us	14.7ms
5 x 5	654us	41.9ms
7 x 7	1310us	83.8ms

Table 2. Spatial convolution times for a bit-serial
SIMD processor with $\emptyset$ = 10MHz

Comparing the data of Table 2 with that of Table 1 reveals a 16 times improvement in performance, bought at the expense of many extra chips. For example, 512 Goodyear MPP or UCL CLIP chips, 256 ICL DAP chips, 128 GEC GRID chips or 64 NTT AAP chips would be required for the implementation of the 64 x 64 pixel patch. In practice, design optimisation has improved the performance of these chips by a factor of 2 - 4. Nevertheless, the question of cost-effectiveness still remains.

Although the data of Table 2 suggest the potential of more than one "sub-image process" per frame-time, the overheads of "patch" load and dump must are no longer negligible. Indeed, a 10MHz pixel-rate would support 64 x 64 pixel "patch" and "sub-image" load/dump times of 0.4ms and 26.2ms respectively. Consequently, such bit-serial SIMD image processors often incorporate local storage and multi-pixel input-output channels to minimise the time penalty of "patch" load and dump.

1.7. SCAPE chip design goals

In contrast to other chip developments for bit-serial SIMD processor implementation, the SCAPE chip is optimised for the implementation of cost-effective rather than high-speed image processing modules (eg. the Brunel IPM). Indeed, the current objective is to fit a 64 x 64 pixel array onto a single PPM printed-circuit-board (for low-cost system implementation) and to support a wide variety of "image processing" algorithms (to attract a higher-volume market), at only a moderately high execution rate. Table 3 puts these design goals into detailed perspective.

Machine	n	#pcbs/ n x n array	Cycle-time	Technlgy	Delivery
UCL CLIP-4	96	96	10000ns	Custom-LSI	1980
ICL DAP	64	256	200ns	MSI/LSI	1980
Goodyear MPP	128	88	100ns	Custom-LSI	1982
GEC GRID	64	16	100ns	Custom-VLSI	1985
Brunel IPM	64	1	100ns	Custom-VLSI	1985

Custom-chip	#pixels	store/ pixel	#trans.	chip area mm^2	line size um	Techn	#pins
CLIP-4	8	32	3k	16	8	nMOS	40
Goodyear MPP	8	32	8k	30	3	CMOS	52
ICL DAP	16	-	10k	40	2	CMOS	64
NTT AAP	64	96	81k	94	3	nMOS	141
GEC GRID-32	32	64	35k	42	2.5	CMOS	88
SCAPE	256	32	150k	56	2.5	CMOS	68

Table 3. Comparison of design goals for bit-serial SIMD image processing systems/chips

The SCAPE chip design followed encouraging results from the Micro-APP project [10] at Brunel University, which aimed to pack a small Associative Parallel Processor on a single VLSI chip. "Real" image processing requirements, provided by British Aerospace and Micro Consultants, enabled the specification of a general-purpose programmable chip architecture to support the widest possible application range. Indeed, associative processing extends the capability of the SCAPE chip well beyond that of its contempories and current research is aimed at exploring the limits of SCAPE chip applications.

2. SCAPE CHIP ARCHITECTURE

The SCAPE chip supports image patches with a "set" of identical cells, each cell providing pixel and "workspace" storage and a bit-serial processing capability. A sequence of global microinstructions is broadcast to all cells for the simultaneous execution of arithmetic, logical or relational operations on all members of a selected "subset" of pixels. The SCAPE allows fully-associative selection of "subsets" and, hence, image processing is performed as a special case of "associative set processing".

2.1. Topology

At first sight, a 2-dimensional cellular array, with the "nearest-neighbour" interconnection topology, seems ideally suited to VLSI. Indeed, when the unit-cell is large compared with its interconnection overheads, an orthogonal array of "near-butting" cells makes very good use of Silicon "real-estate". However, for the small feature-sizes envisaged for VLSI array processing chips, such interconnection networks lead to poor cell packing-densities and high pin-outs, which in turn lead to poor utilisation of Silicon, high power dissipation (due to the extra output-drivers) and expensive chip packaging. Hence, the SCAPE chip architecture is based on a different interconnection topology.

Although the SCAPE chip could be configured "logically" as a 2-dimensional cellular array with orthogonal interconnection, it is "physically" implemented as a "cellular string" with a linear interconnection network. Thus, the cells forming the rows of each image patch are linked end-to-end to form a linear string of cells; each cell being directly connected to the routing channels of the

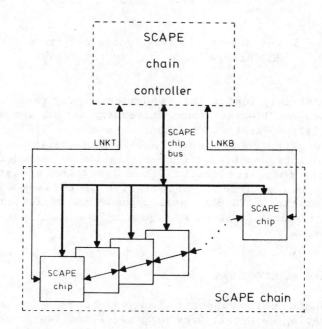

Figure 1. SCAPE-based Patch Processing Module (PPM)

interconnection network. Freed from the "nearest neighbour" connection, the cellular string can be compacted to achieve a high cell packing-density and only the two ends of the string and the routing network contribute to the chip pin-out. Moreover, the simple linking of chips to extend the cellular string, as shown in Fig. 1, allows a high chip packing-density on the PPM printed-circuit-board.

2.2. Associative string processing

The combination of "associative set processing" and "cellular string topology" achieves a versatile medium for the flexible representation of structured data (eg. arrays, tables, trees etc.). Indeed, as an "associative string processor", the SCAPE chip can support a wide range of information processing applications. For example, in the special case of image processing, an "associative string" can be easily reconfigured to support variable patch sizes, flexible pattern matching, list structures and relational tables.

2.3. Image representation

An M_p x N_p pixel "patch", extracted from an M_s x N_s pixel "sub-image" of p-bit pixels, can be processed in a singly-linked string of M_p/r SCAPE chips known as the "SCAPE chain". Each SCAPE chip supports 256 pixels from r rows of the patch.

The value of p can be selected, under program control, according to the needs of the application, with no processing restrictions until p exceeds 8 bits. However, although under program control, the values of M_p and r affect the complexity of the "SCAPE chain", as shown below

#pixels/ chip (r x N_p)	#pixels/ row N_p	#rows/ chip r	patch size	#chips
256	8	32	16 x 16	1
	16	16	32 x 32	4
	32	8	64 x 64	16
	64	4	128 x 128	64
	128	2	256 x 256	256
	256	1		

3. SCAPE CHIP FLOOR-PLAN

The SCAPE chip floor-plan comprises a close-packed square structure of four different (but exactly "butting") functional blocks, as shown in Fig. 2; these are the

AMA: Associative Memory Array comprising 256 words, supporting a 32-bit "data field" and a 5-bit "control field", partitioned into 32 8-word content-addressable memory blocks, each block corresponding to a 8-pixel "row segment" (as indicated in Fig. 3 for 8 32-pixel rows).

BCL: Bit Control Logic selecting 0 or more of the 37 AMA bit-columns in support of arithmetic, logical and relational processing on "declared" fields of activated AMA words. The BCL incorporates index control logic for bit-serial operations and field masking logic. In addition, the BCL also supports a bit-parallel capability, to allow 8-bit and 32-bit processing of activated AMA words.

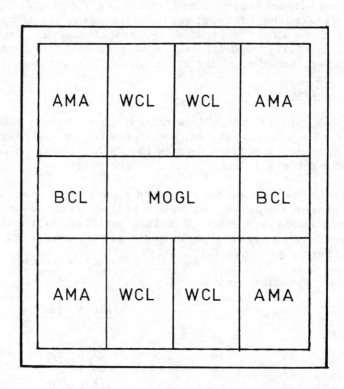

Figure 2. SCAPE chip floor plan

WCL: Word Control Logic activating 0 or more of the 256 AMA words, according to a defined mapping on the response to a content search of selected AMA bit-columns. The WCL incorporates 256 1-bit adder/complementer stages to support bit-serial addition and subtraction operations. Two 256-bit "tag" registers, one to "tag" the currently matching word-rows and the other to "tag" previously matching word-rows, are also included. Activation mappings control the inter-pixel connectivity, required by image processing algorithms, according to the content of the "tag" registers. The WCL is partitioned into 8-word blocks, one for each of the 32 8-pixel "row-segments" in the AMA. These partitions are interconnected by "segment-links", which can be set to configure the "row size" (viz. 0, 8, 16, 32, 64, 128, 256 or more pixels or, indeed, any multiple of 8 pixels) supported by the SCAPE chip. "Row-links" can be "opened" or "closed" such that rows can be processed in "parallel" or "linked" mode. The "chip-links" at the ends of the WCL are used to link the SCAPE chips forming the "SCAPE chain".

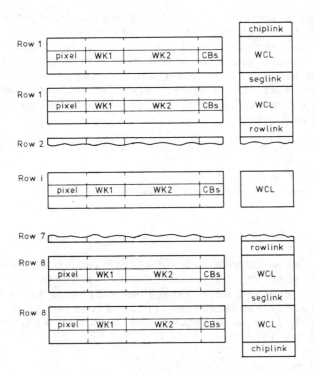

Fig. 3. AMA allocation for 32-pixel patch rows

MOGL: Micro-Order Generation Logic issuing "dynamic micro-orders" to the BCL, WCL and IOMPX derived from the "static micro-order" of the current microinstruction. Input-output multiplexing (to minimise pin-count and package dimensions) and internal clock generation are also performed in the MOGL.

4. AMA ALLOCATION

Up to 4 variable-length fields can be "declared" within the "data field" of the AMA. Of these, up to 3 fields can be used for bit-serial algorithms. Unused fields are "declared" as "blank fields". Thus, the SCAPE chip supports variable-precision processing and "blank fields" provide flexibility in adjusting bit-precision.

Typically, the "data" field is "declared" as a 1 p-bit "image field" and 1 or 2 associated "work fields", with 1 AMA-word allocated to each "pixel" of the "image", as indicated in Fig. 3.

Field-lengths can be re-declared rapidly during algorithm execution, such that "exact" bit-precision can be maintained.

As image processing algorithms proceed, it is likely that the "workspace" requirement per pixel will increase, often with a corresponding decrease in pixel resolution. In such cases, the "workspace" per pixel can be extended in units of 32-bits, by over-writing neighbouring word-rows. Alternatively, more than one word-row can be allocated to each pixel before algorithm execution. In both cases, field "significance" is determined by "control bit" patterns. Thus,

"workspace" per pixel = $(32 - p) + (trunc (256/n) - 1) * 32$ bits

where p = pixel precision
n = number of pixels per SCAPE chip

5. SCAPE OPERATIONS

Bit-serial word-parallel algorithms can be written to perform any "scalar-vector" or "vector-vector" arithmetic, logical or relational operation on binary data patterns. For example, the "vector - vector" addition of those 8-bit operands marked by control-bit B in AMA fields X and Y can be achieved with an algorithm of the type

```
for j := 0 to 7 do
   forall i : B = 1 do Z[i,j] := X[i,j] + Y[i,j]
```

with the corresponding "carries", C[i], set accordingly in the WCL.

Apart from 32-bit load/dump operations, bit-parallel word-parallel "scalar - vector" logical and relational algorithms can also be supported. For example forall i : B = 1 do Y[i] := X and Y[i]   {write ´0´s in X into Y}

forall i : (Y[i].WK1 = X.WK1) and (B = 1) do Y[i].WK2 = X.WK2

where X is an external scalar operand.

6. PATCH LOADING AND DUMPING

Image processing with the "SCAPE chain" has 3 distinct phases, these being "patch" loading, "patch" processing and "patch" dumping. Of these, only "patch" processing gains the full benefit of associative parallel processing, since pixel loading and dumping are essentially sequential operations. Thus, loading and dumping are significant timing overheads. It is for this reason that other SIMD chips (eg. ICL DAP, Goodyear MPP and GEC GRID) provide an input-output data path for each pixel with consequent increase in chip complexity, pin-out and power dissipation. However, such a high input-output bandwidth is impossible with the SCAPE architecture.

To reduce the pixel loading and dumping time penalty, the SCAPE chip utilises a 32-bit read-modify-write operation, which can "exchange" 4 input and output 8-bit pixels in a single clock cycle. Thus, each SCAPE chip can dump 256 "old" 8-bit pixels and load 256 "new" 8-bit pixels in 64 clock cycles. Hence, by using each SCAPE chip bus as an independent "data channel", an entire "patch exchange" could be achieved in 64 cycles.

7. SCAPE CHAIN CONTROLLER

All SCAPE chips in the "SCAPE chain" receive microinstructions and data "broadcast" from the "SCAPE chain controller", as indicated in Fig. 1. This unit comprises the standard bit-slice microprocessor components of typical high-speed microprogram-controllers.

The SCAPE chain controller executes "SCAPE macros", stored as microprograms, can be called from a program running in the host machine. The controller also buffers data transmitted between the "SCAPE chain" and the host.

8. SCAPE SOFTWARE

SCAPE application programs are written entirely in Pascal (or a similar block-structured high-level language). Such programs include calls to "external" pre-compiled procedures, known as "SCAPE macros", which are written with the "SCAPE macro assembly language". Hence, the application programmer selects suitable "SCAPE macros" from the "SCAPE macro library" and only if it is necessary does he resort to "SCAPE macro" creation. In operation, selected "SCAPE macros" are stored as microprograms in the "SCAPE chain controller". Thus, a sequence of "SCAPE macros" is called by the application program, which is suspended during "SCAPE macro" execution.

9. COMPLEXITY AND PERFORMANCE FORECASTS

The SCAPE chip design follows research which involved the specification and evaluation of algorithms and associative processing structures for real-time image analysis. The feasibility of such chip architectures has been investigated in a project which involved the design, fabrication and evaluation of Micro-APP test-chips [10]. More recently, nMOS designs for the major layout blocks of the SCAPE chip floor plan have been evaluated with SPICE analyses. At the time of writing, the SCAPE chip is in advanced stage of design, with most of the CMOS layout blocks completed and much SPICE data available. Hence, the following data can be given with some confidence.

#pixels/chip	= 256
#transistors	= 150K
feature size	= 2.5 micron
technology	= CMOS (p-well)
chip size	= 60 mm^2
clock-rate	= 10 MHz
#pins	= 68

The SCAPE chip will be fabricated by Plessey (Caswell) and first samples are expected in the first quarter of 1985. Hence, until that time, no accurate data on performance can be given. However, the results of recent studies are

8-bit add/subtract fields	1.0us
8-bit add/subtract fileds (signed operands)	1.4us
8 x 8 multiplication (double-precision results)	
scalar - vector	4.4us
scalar - vector (signed operands)	5.6us
vector - vector	15.0us
vector - vector (signed operands)	16.0us

Assuming a SCAPE module comprising a 64 x 64 8-bit pixel array (ie. 16 SCAPE chips) and a 10MHz clock-rate, the following interim forecasts give some idea of SCAPE performance:

Convolution times

	weight precision		
	4-bit	6-bit	8-bit
3 x 3 window	77us	81us	86us
5 x 5 window	233us	264us	289us

Median filter times

3 x 3 window	72us
5 x 5 window	90us

Histogram generation times

	Number of grey levels				
	256	128	64	32	16
Best case	39us	32us	29us	27us	27us
Worst case	422us	416us	412us	411us	410us

Contrast stretching   8.4us

1-pixel binary erosion/dilation   1.5us

10. CONCLUSIONS

The design of the SCAPE (Single-Chip Array Processing Element) chip has been optimised for the cost-effective implementation low-cost microelectronic SIMD (Single-Instruction control of Multiple-Data streams) real-time image processing modules. Indeed, two key features, of the design philosophy of the SCAPE chip, are

(1) the compromise of high-speed cell communication for high pixel packing-density and reduced pin-out to achieve cost-effective use of Silicon

(2) high algorithm versatility for the execution of a wide range of image processing algorithms and, indeed, more general information processing algorithms, since higher-volume marketing also improves the cost-effectiveness of VLSI chips.

Since each "pixel-cell" integrates local memory and processing logic in close proximity, a high intra-cell processing speed can be sustained. Thus, the combination of high packing-density and high-speed routing channels in the "centralised" interconnection network serves to minimise the inequality of column-to-column and row-to-row communication timing. Moreover, the use of fully-associative memory, avoiding the need for address-routing and address-decoders, supports higher algorithm execution-rates and improves the regularity of the chip floor-plan. Hence, the "speed compromise" has achieved a substantial improvement in pixel packing-density without a significant loss in speed. Indeed, SCAPE algorithm execution times are similar or better than those reported for other SIMD "image processing" chips. In fact, comparing SCAPE chip data with the contents of Table 4, indicates this trend.

The high algorithm versatility of the SCAPE chip has yet to be proven in practice. However, SCAPE macro development is prominent in current research projects and the universality of the "associative string processing" structure of the SCAPE chip is strongly indicated. Indeed, current progress in digital (linear) signal processing and relational data base management is most encouraging.

Custom-chip	Memory access (us)	1-bit pixel op. (us)	1-bit neighbour op. (us)	8-bit add (us)	8-bit mult. (us)
CLIP-4	1.6	10	10	80	2400
ICL DAP	>0.1	>0.1	(<1)	1	(<10)
Goodyear MPP	0.1	0.3	1.5	2.5	9
NTT AAP	0.1	0.1	(0.8)	1	–
GEC GRID-32	0.1	0.1	0.6	<1	8

Table 4. Comparison of bit-serial SIMD chip data [11]

11. ACKNOWLEDGEMENTS

The author gratefully acknowledges the enthusiastic contributions of P. Overfield, A. Krikelis and, especially, I.P. Jalowiecki during the preparation of SCAPE performance forecasts and the support of British Aerospace, Micro Consultants, Plessey and the UK "Ministry of Defence" for SCAPE chip development.

12. REFERENCES

1. Lea R.M., A VLSI array processor for image processing, Proc. of Workshop on Algorithmically-specialized Computer Organisations, Purdue University (USA), 1982.

2. Lea R.M., SCAPE: a Single Chip Array Processing Element for image analysis, VLSI 83 (Eds. Anceau F. and Aas E.J, Elsevier Science Publishers B.V. North-Holland), pp. 285 - 294, August 1983.

3. Unger S.H., A computer oriented towards spatial problems, Proc. WJCC, pp. 234 - 239, 1958.

4. Slotnick D.L., Borck W.C., McReynolds R.C., The SOLOMON computer, Proc. AFIPS (FJCC), pp. 97 - 107, 1962.

5. Duff M.J.B., Review of CLIP-4 image processing system, Proc. NCC, pp. 1055 - 1060, 1978.

6. Flanders P.M., Hunt D.J., Reddaway S.F., Parkinson D., Efficient high-speed computing with the Distributed Array Processor, High Speed Computer and Algorithm Organisation (Academic Press (London), 1977), pp. 113 - 127.

7. Batcher K.E., Design of a massively parallel processor, IEEE Trans. Comput., C-29, pp. 836 - 840, 1980.

8. Sudo T., Nakashima T., Aoki M., Kondo T., An LSI adaptive array processor, IEEE-ISSCC, pp. 122, 123 and 307, 1982.

9. McCabe M.M., McCabe A.P.H., Arambepola B., Robinson I.N., Corry A.G., New algorithms and architectures for VLSI, GEC J. of Science & Tech., 48, 2, pp.68 - 75, 1982.

10. Lea R.M., I2L micro-associative-processors, ESSCIRC 79, pp. 104 - 106, 1979.

11. Fountain T.J., A survey of bit-serial array processor circuits, in: Duff M.J.B. (ed.), Computing structures for image processing, pp. 1 - 14, (Academic Press, London, 1983).

VLSI Architectures for Computer Graphics[1]

Gregory D. Abram
Henry Fuchs

Department of Computer Science
University of North Carolina at Chapel Hill, USA

Abstract

Both academic researchers and commercial concerns are increasingly interested in applying VLSI technologies to graphics systems:

• For researchers, graphics systems offer an attractive model for study of computer architectures in VLSI: these systems have a small well-defined set of operations and simple data and control structures, making these systems ripe for applying parallelism and modularization techniques; many of these systems, especially the interactive high-resolution color ones, have severe computation demands that are unfulfilled by solutions embodied in current systems.

• For commercial concerns, there is a rapidly increasing market for interactive graphics systems as personal workstations in which graphics displays replace text-only terminals.

In this paper, we cover: a) the conceptual organization of a "generic" graphics system and its realization in several state-of-the-art commercial products; b) the architecture of several recent VLSI chips and systems and their likely effect on the organization of future graphics systems; c) the architecture of several VLSI-based systems that are currently subjects of research. The design strategies used in these systems -- the structure of parallelism, intertwining of data and computation, the tradeoff between custom and off-the-shelf parts -- may provide insights into other applications as well.

1. INTRODUCTION

The design of graphics systems has been a challenging topic of study for several decades; the demands for ever-increasing performance have always pushed the available technology to its limits. The availability of off-the-shelf TTL circuitry in the early 1970's allowed custom designs of minicomputer-level complexity. The advent of large-scale RAMs allowed systems to store complete images and quickly and randomly address any

---

[1] This research was supported in part by the (USA) National Science Foundation under grant ECS-8300970 and by the Defense Advanced Research Projects Agency contract DAAG 29-83-K-0148.

pixel in them; this capability gave rise to the current boom in color raster systems. Inexpensive microprocessors allowed these frame buffer systems to perform many functions independently of the host computer.

The possibility of custom VLSI promises another level of power in affordable graphics systems. The increased plasticity of custom VLSI allows systems designed using this medium to take on radically different structures than seen heretofore. This paper explores some of these possible structures — a few just recently announced, most yet to come.

We concentrate in this paper on interactive color raster systems aimed at laboratory or office use, mostly for 3D applications; a few related systems that focus on 2D applications are included [Gupta, Sproull, et. al., 1981]. Of course, most of the systems cited can be used for a wide variety of applications, not restricted to 3D. We have intentionally left out systems aimed at the expensive ($1M) flight simulator market, largely due to lack of available information in the public domain [Schachter, 1981], although a number of the systems covered in this paper may be used for flight simulator applications.

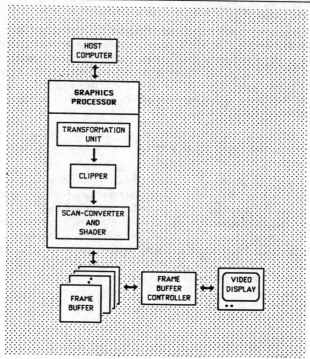

Figure 1: A Typical 3D Raster System Organization

The overall organization of many raster graphic display systems is quite similar (Fig. 1). The central feature is the frame buffer memory in which is stored the image currently being displayed — and perhaps one or more additional images. To relieve the host computer from low-level tasks, one or more processors are attached to most frame buffers. The nature and organization of these processors is one of the major focuses of this paper. Its major tasks for 3D image generation are illustrated in Figure 2. An alternate organization is used for many general purpose workstations, in which the image usually shows one or more pages of mostly textual information (rather than an interactive 3D image). The major tasks involve generation and movement of 2D image data; the typical hardware organization for such systems is shown in Figure 3.

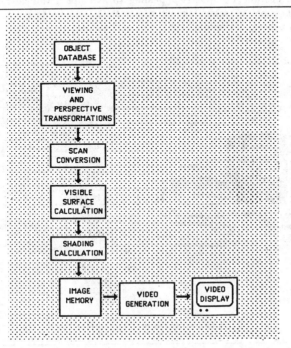

Figure 2: Functional Steps for 3D Image Synthesis

In studying many graphic system designs currently being developed, several distinct strategies become evident.

- Implementing innermost loops in hardware

One obviously reasonable strategy to consider is to transfer an often-executed inner loop from software to hardware [Atwood, 1984; T. Ikedo, 1984]. The details of these designs are discussed in section 2 . As some have noted, however, this strategy may not always succeed [Pike, 1984]. The new hardware inner loop solver may add so much more overhead as to swamp any gains it produces in solving the inner loop faster.

- Integrating a boardful of functions onto a single chip

Some systems have succeeded by restructuring an extant solution into VLSI components and thereby reducing a module that formerly needed one or more boards of parts to a few custom chips [Clark, 1982]. As will be seen below, however, it is not always obvious how to restructure the board-level function in such a way as to enable a VLSI-based solution.

- Alternative architectures

The restructuring for a VLSI-based solution can extend beyond the board to system level; with custom VLSI, it is appealing to attempt a radical restructuring of the problem in hopes of achieving a solution that's much more attractive in this new medium [Fuchs and Poulton, 1981; Fuchs, Poulton, et. al., 1982; Kedem and Ellis, 1984].

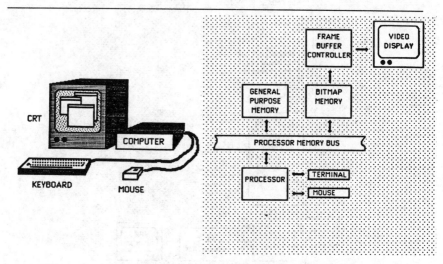

Figure 3: A Bitmap Display and Typical Hardware Organization

In this paper, we review several current and several proposed systems which take advantage of VLSI technology. Associated with each will be a figure noting the structure of such a system. It should be kept in mind that these architectural layouts are conceptual models only, reflecting our own understandings, and may bear only superficial resemblance to actual implementations.

2. HARDWARE FOR CRITICAL LOW-LEVEL FUNCTIONS

One characteristic of computer graphics systems is that a few low-level functions are used extremely frequently and, therefore, account for large portions of the total work done. Much work has been done to speed up the software algorithms used to perform these functions. Major advancements may be achieved by supporting these functions in hardware. Two such functions are the line drawing algorithm, which determines which pixels best approximate a line, and the "raster op", a complex function which allows logical functions between arbitrary rectangular regions of a bitmap display [Bechtolsheim and Baskett, 1980; Thacker and McCreight, 1979].

2.1 VLSI Support for Line Drawing

Although major strides have been made in the design of raster graphics systems, random-scan (also called vector or calligraphic) systems remain the technology of choice for line-drawing applications. This is for two reasons: image generation time and image quality. Raster systems must compute the set of pixels which best approximate lines and set them accordingly; random-scan systems use analog circuitry to drag the electron beam across the CRT screen from endpoint to endpoint. Since raster systems have only a relatively coarse grid of addressable pixels, images show distracting staircasing effects along edges (unless costly anti-aliasing algorithms are used); in contrast, lines on random-scan systems are smooth.

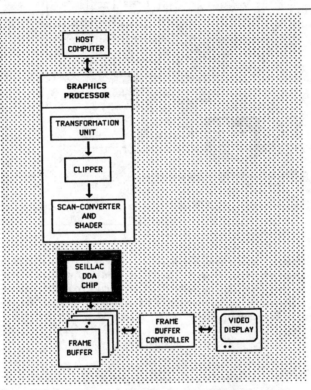

Figure 4: Organization of System Using SEILLAC DDA Chip

The SEILLAC-7, a new graphics system built by the Seillac Co., Ltd., utilizes a custom ECL DDA chip to achieve extremely high line drawing rates (Fig. 4). It is claimed to be about five times faster than previous raster systems that lack such special-purpose hardware (though other systems claim similar speeds, such as the Ramtek 2020) [Ikedo, 1984]. This chip, which achieves a speed of about 40 nanoseconds per pixel in the line, includes a function to modulate the pixel intensity to alleviate the staircasing effects. In doing so, the images generated are claimed to approach stroke-drawn systems both in image quality and in vector drawing speeds.

2.2 Bitmap Manipulations

A recent development in professional workstations has replaced the standard ASCII terminal with a high resolution black-and-white frame buffer system (a "bitmap" display).
This approach, pioneered in the Xerox Alto system in the early 1970's [Thacker,
McCreight, et. al., 1971], offers many advantages over standard ASCII terminals; for
example, high quality graphics and arbitrary fonts can be used for document preparation.

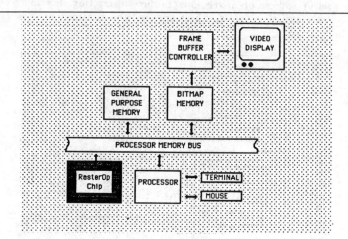

Figure 5: Bitmap System Organization Using RasterOp Chip

Bitmap displays require that the graphics system perform operations (Raster Ops)
on bitmap memories efficiently (Fig. 5). Scrolling a bitmap window requires that the window be copied up one line; this must be done very quickly if the display is to be useful.
Raster Ops typically allow logical operations on source and destination bitmaps; the copying process uses the function:

$$f(source, destination) = source$$

whereas the function:

$$f(font, destination) = \sim font$$

may be used to write reverse video characters.

The implementation of the RasterOp function, however, is tricky. First, source and
destination areas may overlap; the algorithm must to careful to operate in an order which
ensures that data will not be overwritten before it is used. For example, if the destination is to the *left* of the source, the operation must proceed from left to right across the
source, whereas if the destination is to the *right* of the source, the pass must be in the
opposite direction.

The problem is further complicated by the organization of the bitmap memory.
Bitmap displays are often organized with 16 or 32 horizontally adjacent pixels in a single
word. Since regions do not necessarily fall on word boundaries, corresponding pixels in

source and destination words may fall at different bit positions within the words. In order to operate on the several pixels within each memory word in parallel, *two* source words must be available to be to aligned with the data within the destination data word. The logical operation is then applied to the aligned words, and the result written to the destination location. This must be repeated for each word which contains a destination pixel.

Silicon Compilers, Inc., in conjunction with Sun Microsystems, Inc., have implemented a chip to support the RasterOp function [Iannamico and Atwood, 1984]. This chip utilizes a two word FIFO to manage the source data words; the adjacent words are fed to a barrel shifter to align them with the destination data. Alternately, a pattern register is available for repetitive source data (for example, if a background pattern is to be written). The data words are fed to a simple ALU to compute the logical operation; a function decoder allows the host to choose among 8 possible functions of pattern, source, and destination pixel values. Finally, mask registers can be used to protect bits of the destination data words which lie outside the destination area.

A substantial amount of work remains for an external controller to do (either the host CPU or an external finite state machine). Unlike the Seillac DDA chip outlined above, the looping here must be handled by the controller. The chip has no memory-addressing capability; it must rely on the controller to spoon-feed it input data and to return the results to the destination memory locations. The chip does, however, provide functions which may be costly for conventional microprocessors, including in particular the arbitrary 32-bit shift necessary for data alignment.

3. INTEGRATED SYSTEM COMPONENTS

Whereas greater performance can be had by supporting critical functions in hardware, both performance and cost can be addressed by integrating large parts of the conventional graphics system. Some functions typically built out of large numbers of chips can, in fact, be implemented directly on a single (or a very few) VLSI ICs. This substantially decreases the chip count and can greatly improve the performance.

3.1 TI 4161 Memory Chip

Two related problems plague the frame buffer memory designers: 1) contention between image generation and scan-out for memory access, and 2) the high part count (and associated cost) of satisfactory designs (Fig. 6). For a 1024x1024x1 system refreshed at 60 Hz., a pixel (ie. one bit of the frame buffer memory) must be available for display every 16 nanoseconds (or less). This rate can be achieved by interleaving the pixel memory among several memory chips, which are read in parallel into a high speed shift register which then shifts pixels out at the desired rate. Assuming that the frame buffer is built using 64Kx1 RAM chips, 16 chips are required for a 1024x1024 bitplane, and scanout requires a memory cycle every 16x16 = 256 nanoseconds, leaving little for image generation (unless very high speed--and therefore expensive--memories are used). Using

Possible Memory Organizations		
memory chip size	desired access rate	data path width
256Kx1	64nsec.	4
64Kx1	256nsec.	16
16Kx4	1024nsec.	64

16Kx4 RAMs, we use the same number of chips and get 64 pixels in parallel, and we need a memory cycle every 1024 nanoseconds. Unfortunately, we now need four times as many memory chips and the data path is four times as wide. (Also, with these 4-bit wide chips, modifying a single bit is often awkward, necessitating a read-modify-write operation.)

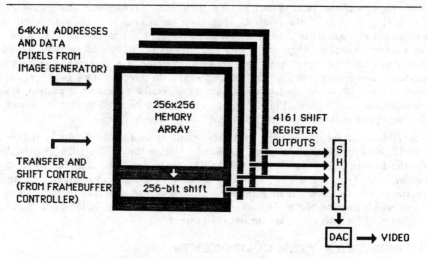

Figure 6: A Possible Frame Buffer Design Using TI 4161 RAMs

In other words, achieving the necessary data rates requires using small-capacity RAMs; whereas achieving low parts counts requires using large-capacity RAMs.

Texas Instruments has recently brought to market a special dynamic memory chip to help solve this problem [Pinkham, Novak, et. al., 1983]. Much like conventional 64K RAMs, the TI 4161 memory is organized as 256 rows of 256 columns. The difference is that a 256-bit 40 nanosecond shift register is included. A command causes an *entire row* to be transferred to the shift register; the chip then acts as two completely independent chips: the 256-bit shift register and a normal 64Kx1 DRAM. In this manner, the chip allows a low system parts count while not tying up memory for scan-out (one memory cycle accesses 256 pixels).

3.2 The Geometry Engine

Three-dimensional image generation requires that each coordinate in the scene be transformed from a object space coordinate system to an image space system and then clipped to the visible region. Because these functions are time-consuming (including multiplies and divides) and lie in the critical data path, this was one of the first image generation components to be implemented directly in hardware.

James Clark, of Stanford University and Silicon Graphics, Inc., has implemented a chip (the Geometry Engine) which, when organized in a twelve-stage pipeline, performs three-dimensional viewing transformations, a perspective transformation and clipping (Fig. 7) [Clark, 1982]. It achieves a rate of 65,000 coordinate points per second and (unlike most earlier systems) operates on floating point numbers. Because a relatively small number of identical ICs are used to implement this pipeline, the cost is low, The cost will be lower still when multiple copies of the present IC will fit onto a single die.

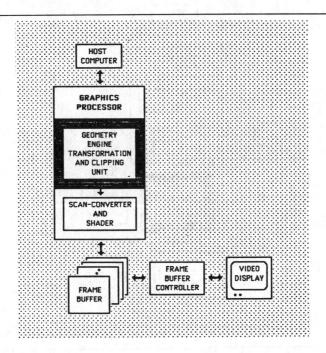

Figure 7: System with Geometry Engine for Transformation and Clipping

3.3 Graphics Display Controllers

Possibly the first specialized integrated circuits for video generation were single-chip video sync generators. Even with such chips, graphics display controllers typically require large amounts of logic, and hence are quite expensive to build out of off-the-shelf MSI and SSI components. Lately, however, two VLSI graphics display controllers have entered the market, each designed for a specific corner of the graphics market.

The NEC 7220 (second sourced as the Intel 82720) is designed to handle high-resolution (1024x1024) color raster graphics systems (such as the Vectrix VX384, a 670x480 system with 9 bits per pixel). The 7220 sits between the host processor (often an Intel micro) and the video memory (Fig. 8). Its video generation circuitry provides a great deal of flexibility, including provisions for zooming, panning, and windowing the image, plus the ability to use a light-pen input device. It also supports image generation by on-board line, arc, area fill and other graphic primitive display functions. Using the 7220, a complete high quality graphics system can be added to a microprocessor system at little more than the cost of the memories and the controller itself.

In contrast, the TI TMS9118 family of graphics display controllers are aimed at the low-cost world of video games, requiring only three chips to add graphics capability onto a standard microprocessor (Fig. 9) [Williamson and Rickert, 1983]. Although they allow only low resolution 256x192 images, they contain support for several specialized functions,

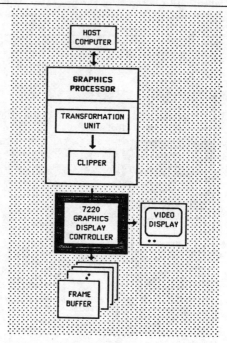

Figure 8: Graphics System Using 7220/82750 Graphics Display Controller

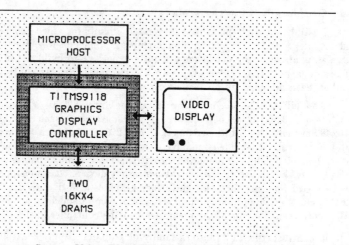

Figure 9: Low-cost System Using TI TMS9118 Graphics Display Controller

including 32 so-called "sprites". A sprite is a small object defined by a rectangular grid of pixels whose position on the screen can be set by simply storing its location in a register, rather than actually copying its pixels from place to place in a (full-image) frame buffer. By using sprites for moving objects in the display, even extremely low-cost devices can support certain classes of very high quality interaction.

4. ALTERNATIVE ARCHITECTURES FOR VLSI

Several current research projects are investigating ways to restructure the traditional graphics architecture to take better advantage of VLSI technology; in particular, the capability of applying potentially many specialized processors to the problems of image generation. These alternative architectures divide into two classes: those that divide the problem in image space and those that divide the problem in object space. Image space strategies divide the *image plane* into independent subsets and associate a separate processor to each. Object space strategies instead divide the *object database* and assign a processor to each.

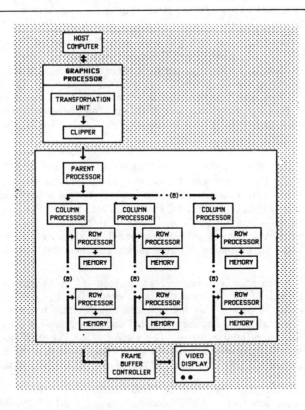

Figure 10: Clark and Hannah 8x8 Display Architecture

4.1 Image Space Strategies

A hard constraint on the performance of raster systems is the bandwidth into the frame buffer memories. However, we can increase the bandwidth to the memories by splitting the image memory into separate components (on the image generation side; the memory still should look contiguous to the scan-out hardware). By associating pixel generating power with each component, these separate components can be accessed in parallel, effectively multiplying the bandwidth to the memories by the number of separate components. In this section, we look at two such strategies. (For speeding up the restricted case of images composed solely of axis-oriented, filled rectangles, see [Whelan, 1982].)

4.1.1 Clark and Hannah

James Clark and Marc Hannah have proposed an image space segmentation approach which splits the image screen into segments the size of the RAM chips used (Fig. 10) [Clark and Hannah, 1980]. This organization is similar to an earlier system in [Fuchs and Johnson, 1979]. Both these systems distribute the image buffer in an interlaced fashion in X and Y among many small memories, each controlled by a small processor. For example, a 1024x1024 bit plane can be built out of 64 16K RAMs. These RAMs are interlaced so that for any 8x8 area of the bit plane, one bit comes from each of the these systems RAM chips. Thus, each memory contains every eighth pixel in every eighth row.

Clark and Hannah's system contains an intermediate layer of "column" processors between the main "parent" processor and the memory controlling "row" processors. Each of these processors then does a share of the image generation computation. To generate a line, the parent processor first determines the starting column, slope, line width and ending column of the line, and transfers this information to the column processors. The column processors then determine the part of the line intersecting the associated column of the image memory and transfers this to the row processors. Finally, the row processors actually write pixels into the image memory.

4.1.2 Pixel-Planes

We, together with colleagues A. Paeth and A. Bell at Xerox Palo Alto Research Center, have been working on an image-generating system, "Pixel-planes" that performs low level pixel operations within "smart" custom memory chips that make up the frame buffer (Fig. 11) [Fuchs and Poulton, 1981; Fuchs, Poulton, et. al., 1982]. The memory chips autonomously perform, 1) scan conversion (calculating the pixels that fall within a line-segment, convex polygon, or circle), 2) visibility calculations based on the depth ("Z") buffer algorithm, and 3) pixel painting (either "flat" or a limited Gouraud smooth shading).

Efficient implementation is possible because each of the above operations can be performed by variations of the same calculation at every pixel, $F(x,y)=Ax+By+C$ where x,y is the address of the pixel. This function can be efficiently realized on silicon by a complete binary tree with a pixel at each terminal node and a one-bit adder paired with a one-bit delay at each non-terminal node. This circuitry and the other needed processing circuitry (a one-bit ALU at each pixel) is sufficiently compact so that the area of the chips consist of half standard memory cells and half the processing circuitry described above.

Since both shading and depth can be formulated in similar equations, Pixel-Planes based systems can perform Gouraud-like smooth shading and Z-buffer visible surface computations. Two working prototypes have been built at UNC; the latest prototype's chips each contain 2K bits of memory distributed among 64 pixel processors, each with 32 bits of memory. Based on conservative speed estimates, (10 MHz clock), the system is expected to process 25,000 to 30,000 arbitrarily-sized polygons per second.

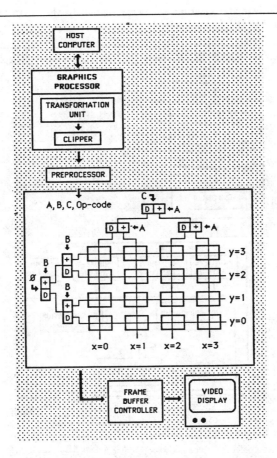

Figure 11: A Pixel-Planes System (4x4 resolution)

4.2 Object Space Subdivision Approaches

An alternative opportunity for parallel processing in image generation is to subdivide the *input data*, assigning separate hardware to each subdivision. Some of the earliest real-time flight simulation systems used this approach; unfortunately, at the time hardware had to be built out of a large number of simple parts, and was therefore extremely expensive and limited in scope [Schumacker, Brand, et. al., 1969]. Using VLSI technology, however, small, specialized processors can be built to perform the necessary operations. Several new designs have been proposed along these lines.

4.2.1 Gershon Kedem's CSG Machine Gershon Kedem has proposed an architecture for the display of objects defined using Constructive Solid Geometry (CSG) (Fig. 12) [Kedem and Ellis, 1984]. CSG is a strategy for computer-aided design in which designs consist of several primitive shapes (spheres, cones, prisms etc.) which are combined using regularized set operations (UNION, INTERSECTION, ADDITION and SUBTRACTION). CSG structures are very naturally represented as binary trees in which leaf nodes correspond to primitives and internal nodes correspond to the operation which combines the objects described in the two subtrees.

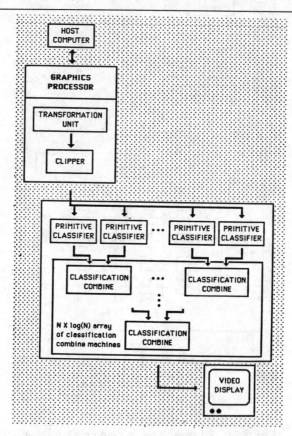

Figure 12: Kedem's CSG Machine Architecture

Kedem's approach instantiates the CSG tree directly in hardware. A reconfigurable tree structure is built which consists of two types of nodes: Primitive Classifiers (PCs), for leaf nodes, and Combine Classifiers (CCs), for internal nodes. To compute a pixel value, the PCs compute (in parallel) the intersections of the ray rooted at the eye point and passing through the pixel center with their associated primitive objects. These intersections (actually line segments of the ray) filter up the tree. Each CC takes the line segments of its left and right subchild and applies its operator on them and passes the result up the tree. The final result, produced at the root of the tree, is then used to compute a pixel shade.

4.2.2 Cohen and Demetrescu Cohen and Demetrescu, in [Cohen and Demetrescu, 1980] have proposed a system that assigns a processor to each potentially visible polygon in the image space (i.e., already transformed world model polygon) (Fig. 13). These, processors are connected as a pipeline and are operated in synchrony with the video generation. For each pixel on the screen, a token is passed through the pipeline of polygon processors. This token carries the shade and depth of the closest point found for this pixel. This depth is the distance from the viewing position of the closest polygon encountered at this pixel; thus the shade is the best guess so far of the color seen at this pixel. Each processor in turn tests whether the pixel lies inside its polygon. If the point lies inside, the processor compares this depth with its polygon's depth at this point. If the polygon's depth is closer, its depth and color replace the token's data. For real-time image generation, tokens pass in raster-scan order and travel at video rates; that is, each processor must make each decision in one pixel time.

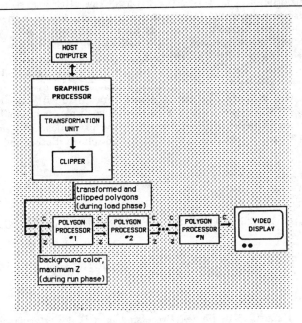

Figure 13: Cohen and Demetrescu's Pipelined Architecture

An elegant feature of this approach is that the pixels stream out of the end of the pipeline in raster-scan order and each value represents the color of the nearest polygon at that pixel; thus the data can be routed directly onto a video display screen.

Weinberg, in [Weinberg, 1981], proposes an elaboration on this design which addresses the problem of anti-aliasing by passing multiple depth-sorted tokens for each pixel along with subpixel masks. Each processor then determines the portion of the pixel covered by its polygon, and compares with the token's mask when the correct position in the depth order is found. If preceding polygons do not completely obscure it, it is added to the token chain. Subsequent tokens are then examined to see if the new polygon completely obscures them, deleting those that are. A filter section at the output uses this data to determine an output shade for each pixel.

This general approach features great modularity; it consists of identical processors hung together in a simple pipeline. It is easily expandable by simply adding more processor chips. The design costs are held down by the fact that only a single IC needs to be developed; manufacturing costs are held down by the simple structure. The only difficulties may be 1) implementing enough processors so that there is one for each and every polygon in the most complex scene in the intended application, and 2) making each processor sufficiently fast to complete all its calculations for a pixel in one pixel time.

5. SUMMARY AND CONCLUSIONS

As the reader is likely to gather from the above list of designs, we are currently witnessing a blossoming of creative designs for harnessing a new medium to solve an old problem. The good news in all this is that with all this attention, there is likely to be substantial progress; indeed, virtually with each passing month a new system with increased performance is introduced --usually found to contain some custom integrated circuitry. In the next few years, many of the designs described above will, no doubt, be developed, refined, and tested. The effective ones will be adopted, the others will be improved or abandoned. Further in the future, we may see designs that integrate even more of the display system functions -- perhaps including the display itself within the processing and image memory. We can then look forward to carrying around a display the size of a book, whose surface is a high-resolution display with built-in high-speed image generating capabilities -- thus approaching the predictions of visionaries' "dynabooks" [Kay, 1977] and eye-glass mounted "ultimate displays" [Sutherland, 1965].

References

Atwood, J. "Raster-Op Chip Overview", Silicon Compilers, Inc., Los Gatos, California (1984).

Bechtolsheim, A. and F. Baskett, "High-Performance Raster Graphics for Microcomputer Systems", *Computer Graphics (SIGGRAPH '80 Proceedings)*, Vol. 14, No. 3, (July 1980) 43-47.

Clark, J. and M. Hannah, "Distributed Processing in a High-Performance Smart Image Memory", *VLSI Design*, Vol. I, No. 3, 4th. Quarter, (1980).

Clark, J. "The Geometry Engine: A VLSI Geometry System for Graphics", *Computer Graphics (SIGGRAPH '82 Proceedings)*, Vol. 16, No. 3, (July 1982) 127-133.

Cohen, D. and S. Demetrescu, Presentation at SIGGRAPH '80 Panel on Trends on High Performance Graphic Systems, (1980).

Fuchs, H. and J. Poulton, "PIXEL-PLANES: A VLSI-Oriented Design for a Raster Graphics Engine", *VLSI Design*, Vol. II, No. 3, 3rd. Quarter, (1981).

Fuchs, H. and B. W. Johnson, "An Expandable Multiprocessor Architecture for Video Graphics", *Proceedings of 6th Annual (ACM-IEEE) Symposium on Computer Architecture*, (April 1979) 58-67.

Fuchs, H., J. Poulton, A. Paeth and A. Bell, "Developing Pixel-Planes, A Smart Memory-Based Raster Graphics System", *Proc. MIT Conference On Advanced Research in VLSI,* Artech House, Dedham, MA., (January 1982).

Gupta, S., R. Sproull, I. E. Sutherland, "A VLSI Architecture for Updating Raster-Scan Displays", *Computer Graphics (SIGGRAPH '81 Proceedings),* Vol. 15, No. 3, (August 1981) 71-78.

Ikedo, T., "High-Speed Techniques for a 3-D Color Graphics Terminal", *IEEE Computer Graphics and Applications,* Vol. 4, No. 5 (1984).

Kay, A., "Microelectronics and the Personal Computer", *Scientific American,* Vol. 237 No. 3, (September, 1977).

Kedem, G. and J. Ellis, "Computer Structures for Curve-Solid Classification in Geometric Modelling", Technical Report TR137, Department of Computer Science, University of Rochester, (May, 1984).

Pike, R., Presentation at University of North Carolina at Chapel Hill, (1984).

Pinkham, R., M. Novak and K. Guttag, "Video RAM Excels At Fast Graphics", *Electronic Design,* (July 21, 1983) 161-172.

Schachter, B., "Computer Image Generation for Flight Simulation", *IEEE Computer Graphics and Applications,* Vol. 1, No. 4, (1981).

Schumacker, R. A., B. Brand, M. Gilland, W. Sharp, "Study for Applying Computer-generated Images to Visual Simulation", *U. S. Air Force Human Resources Lab. Tech. Rep.* AFHRL-TR-69-14, (September 1969).

Sutherland, I., "The Ultimate Display", *Proceedings of the IFIP Congress,* Vol. 2, 1965.

Thacker, C. P., E. M. McCreight, B. W. Lampson, R. F. Sproull, D. R. Boggs, "ALTO: A Personal Computer", Xerox Corp., (1979) in Siewiorek, D. P., C. G. Bell, and A. Newell, *Computer Structures: Principles and Examples,* McGraw-Hill, (1982) 549-572.

Weinberg, R., "Parallel Processing Image Synthesis and Anti-Aliasing", *Computer Graphics (SIGGRAPH '81 Proceedings),* Vol. 15, No. 3, (August 1981) 55-61.

Whelan, D., "A Rectangular Area Filling Display System Architecture", *Computer Graphics, (SIGGRAPH '82 Proceedings),* Vol. 17, No. 3, (July 1982) 147-153.

Williamson R. and P. Rickert, "Dedicated Processor Shrinks Graphics Systems to Three Chips", *Electronic Design,* (August 4, 1983) 143-148.

PART III:
VLSI ARCHITECTURES FOR SYSTOLIC ARRAYS

Experience with the CMU Programmable Systolic Chip

Allan L. Fisher, H. T. Kung, and Kenneth Sarocky

Department of Computer Science, Carnegie-Mellon University
Pittsburgh, Pennsylvania 15213

Abstract

The CMU *programmable systolic chip* (PSC) is an experimental, microprogrammable chip designed for the efficient implementation of a variety of systolic arrays. The PSC has been designed, fabricated, and tested. The chip has about 25,000 transistors, uses 74 pins, and was fabricated through MOSIS, the DARPA silicon broker, using a 4 micron nMOS process. A modest demonstration system involving nine PSCs is currently running. Larger demonstrations are ready to be brought up when additional working chips are acquired.

The development of the PSC, from initial concept to a silicon layout, took slightly less than a year, but testing, fabrication, and system demonstration took an additional year. This paper reviews the PSC, describes the PSC demonstration system, and discusses some of the lessons learned from the PSC project.

Introduction

Using massive parallelism and pipelining, the systolic array concept[1] allows a system implementor to design extremely efficient machines for specific computations. But for some applications such as computer vision that call for hundreds of subroutines to be used routinely, it is impractical to produce a new systolic array processor for each subroutine. In this case, *programmable* systolic array processors must be used to provide the required flexibility.

However, to make a processor programmable takes additional hardware. This concern is especially significant for systolic arrays, as their performance relies on the use of large num-·bers of cells in the array. To be cost-effective, each cell should use as few chips as possible.

The purpose of the PSC project has been to study the feasibility and issues of implement-ing a cell (for a variety of systolic arrays) with one *single, programmable* chip, as depicted in Figure 1. A particular systolic cell can be implemented by microprogramming a PSC, and many PSCs can be connected at the board level to build systolic arrays of many different types and sizes.

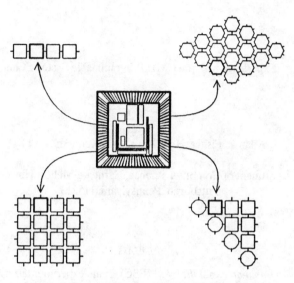

Figure 1. PSC: a building-block chip for a variety of systolic arrays.

The PSC is perhaps one of the first microprogrammable chips designed to be used in large groups. Besides being an architectural experiment, the PSC project also represents a major chip design experiment in a university environment. Prior to the PSC project, CMU had no experience in designing chips of this scale. The experience resulting from the project, with respect to both the architecture and design of the PSC, has been invaluable.

This paper reports some of these experiences, describes the current PSC demonstration system, and explains how the PSC implements the systolic array in the demonstration system to perform convolution or filtering operations. In the next section, we first give a brief overview of the PSC. Detailed descriptions of the PSC architecture and design have been reported in other papers.[2, 3, 4]

PSC: A Programmable Systolic Chip

The PSC project started in October of 1981. In order to ensure sufficient flexibility to cover a broad range of applications and algorithms, we chose at that time an initial set of target applications for the PSC to support, including signal and image processing, error correcting codes, and disk sorting. The demands of these applications have resulted in the following design features:

- 3 eight-bit data input ports and 3 eight-bit data output ports.

- 3 one-bit control input ports and 3 one-bit control output ports.

- Eight-bit ALU with support for multiple precision and modulo 257 arithmetic.

- Multiplier-accumulator (MAC) with eight-bit operands and 16-bit accumulator.

- 64-word by 60-bit writable control store.

- 64-word by 9-bit register file.

- Three 9-bit on-chip buses.

- Stack-based microsequencer.

Note that no conventional, commercially available microprocessor components fulfill the needs of such a programmable systolic chip. Unlike the PSC, conventional microprocessors do not have fast, on-chip multiplier-accumulator circuits which are crucial for high-speed signal and image processing, they do not have enough off-chip I/O bandwidth and on-chip bus bandwidth to pass data from chip to chip with a speed sufficient to balance the computation speed, they are not equipped with I/O ports for passing "systolic control bits," they are not suited for the modular arithmetic needed in applications such as error-correction, and they usually do not have on-chip RAM for program memory. A number of more specialized processors having some of these features have appeared in the past several years, but none has all of them.

With optimized circuit and layout designs, the PSC should operate at a cycle time of no more than 200 ns, although the prototype PSCs we now have are found to be three to eight times slower. Reasons for this are discussed below. Assuming a 200 ns period, microcode examples indicate the following performances:

- A decoder of Reed-Solomon error-correction codes[5, 6] that can correct up to 16 erroneous bytes in 256-byte blocks can be implemented with 112 PSCs with a throughput of 8 Mbits/second. Encoding at the same rate can be achieved with only 32 chips. The fastest existing decoder of which we are aware operates at 1 Mbit/second and uses 500 chips.

- A digital filter (FIR or IIR) with eight-bit data and coefficients and k taps can be computed with k PSCs, taking one sample each 200 ns. For a 40 tap filter, this amounts to 400 million operations per second (MOPS), counting each inner product step (eight-bit multiply, 16-bit add) as two operations. This is equivalent to 600 MOPS for pure eight-bit arithmetic.

- For applications requiring more accuracy, a filter with 16-bit data and eight-bit coefficients and m taps can be computed with m PSCs, taking one sample each 1.2 μs. Thus with 40 PSCs, a 40 tap filter can be computed at a rate of 67 MOPS, counting each inner product step as two operations. This is equivalent to 200 MOPS for eight-bit arithmetic.

- A disk sorter implemented with 17 PSC chips and 16 Mbytes buffer memory can achieve an order of magnitude of speed-up over conventional minicomputers.

Use of the PSC in Implementing Systolic Arrays for Convolutions

The current demonstration system for the PSC performs two-dimensional (2-D) convolutions using general 3×3 kernels. In this section we describe briefly how the PSC is used to implement a systolic array for 2-D convolutions.

212

Mathematically, the 2-D convolution problem with a $k \times k$ kernel is defined as follows:

Given the 2-D kernel of weights $w_{ij}, i = 1, 2, \ldots, k, j = 1, 2, \ldots, p$, and the 2-D input image $x_{ij}, i = 1, 2, \ldots, m, j = 1, 2, \ldots, n$, with $k \ll m$ and $p \ll n$,

compute the output image $y_{rs}, r = 1, 2, \ldots, m - k + 1, s = 1, 2, \ldots, n - p + 1$, defined by

$$y_{rs} = \sum_{i=0}^{k-1} \sum_{j=0}^{p-1} w_{i+1, j+1} \cdot x_{i+r, j+s}.$$

The 2-D convolution problem is one of the most computation-intensive tasks in signal and image processing. For example, a 2-D convolution using a general 9×9 kernel requires 81 multiplications and 80 additions to generate each pixel in the output image.

1-D Convolution Implementation

We first illustrate a systolic array design for the one-dimensional (1-D) convolution problem, which is simpler than the 2-D one. The 1-D problem is defined as follows:

Given the sequence of weights $\{w_1, w_2, \ldots, w_k\}$, and the input sequence $\{x_1, x_2, \ldots, x_n\}$,

compute the result sequence $\{y_1, y_2, \ldots, y_{n+1-k}\}$ defined by

$$y_i = w_1 x_i + w_2 x_{i+1} + \cdots + w_k x_{i+k-1}.$$

Figure 2 depicts one of the well-known systolic arrays[1] for the case $k=3$.

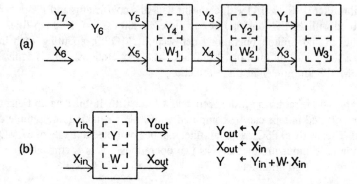

Figure 2. (a) Systolic array for 1-D convolutions, and (b) its cell definition.

We can program the PSC to implement each of the systolic cells; the program takes only one instruction to implement all the operations depicted in Figure 2(b). After an initialization phase in which the weights are loaded, the inner loop of the algorithm uses one PSC microinstruction, coded as follows:

```
Bus1=Sda, Bus2=Sdb, Pus3=Lo,
SdaOut=Val3, SdbOut=Val2,
MacX=Hold, MacY=Val2, MacZ=Val1, MacOp=AddZ,
Jump=OnCCO, CCO=Sca, ScaOut=Pass.
```

The lines of this microinstruction have the following effects, all in a single cycle:

1. Bus 1 carries Y_{in}, read from systolic data port A, bus 2 carries X_{in}, read from port B, and bus 3 carries Y_{out} from the previous operation, available as the output of the MAC.

2. Output port A receives Y_{out} from the previous operation (the value on bus 3), and port B receives $X_{out} = X_{in}$.

3. The MAC holds the cell's weight W in its x register, sets its y register to X_{in} (the value on bus 2), and sets its z register to Y_{in} (the value on bus 1). It then computes $x \cdot y + z$, or $W \cdot X_{in} + Y_{in}$. This value will be sent to output port A during the next cycle.

4. The instruction loops in place until systolic control signal A arrives, meaning it is time to reinitialize. Control then passes to the next instruction, and the control bit is sent on to the neighboring cell.

2-D Convolution Implementation

The above systolic array design for 1-D convolutions can be generalized to designs for 2-D convolutions. In particular, we will use a linear systolic array of k^2 cells to perform 2-D convolutions using $k \times k$ kernels. This systolic array will have the nice "scalable" property that its interface with the outside world is independent of k. That is, when the kernel size is increased, we need only expand the linear array accordingly, without changing its I/O interface. There exist at least two such "scalable" systolic array designs for the 2-D convolution problem, one requiring a memory associated with each systolic cell to buffer one line of image,[7] and the other one requiring no such memory for each cell.[8] For implementation simplicity we use the latter one.

The input image x_{ij} is fed to the systolic array in columns $2k-1$ pixels high, and the output image y_{ij} is generated by the systolic array in swaths which are k pixels high. As depicted in Figure 3, pixels from the input image enter the systolic array in two x-data streams—the x_{ij} with odd j come in with the top x-data stream, and the x_{ij} with even j come in with the bottom x-data stream. In each cycle, a cell will choose a value from one of the two x-data streams to multiply by the weight stored in that cell. The simple rule is that a cell should use values from one x-data stream for k consecutive cycles before switching to the other stream; and continue alternating in this manner. By utilizing the systolic control ports provided by the PSC, a control signal can be sent conveniently from cell to cell to signal the switching from one x-data stream to the other for each cell. It takes no more than two PSC instructions to implement all the cell operations depicted in Figure 3. This implies for example that 2-D convolutions with 9×9 kernels can be implemented by 81 linearly connected PSCs, capable of producing one output pixel every 400 ns assuming a state-of-the-art design.

214

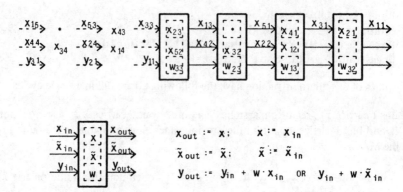

Figure 3.   Linear systolic array for 2-D convolutions and its cell definition.

For large kernels, it is necessary to save partial results y_i in double or higher precision to ensure numerical accuracy of the computed results. As for the case of systolic arrays for 1-D convolution,[1] this can be achieved most cost-effectively by using a dual of the design in Figure 3, where partial results y_i stay in cells but the x_i and w_{ij} move from cell to cell in the same direction but at two speeds. With this dual design, high-precision accumulation can be effectively achieved by the on-chip multiplier-accumulator circuit, and the number of bits to be transferred between cells is minimized. We of course still need to transfer the final computed values of the y_i out of the array, but they, being the final results, can be truncated and transferred in single precision. It is generally preferable to have the w_{ij} rather than the x_i going through an additional register in each cell, since there are two x-data streams but only one weight stream. In fact the communication cost for the weight stream can be totally eliminated if the register file of the PSC is large enough to hold a complete weight table.

Note that in generating adjacent output swaths, some input pixels are fed into the systolic array twice. To avoid having to bring these pixels out from memory twice, a cache that can hold $k-1$ lines of input pixels can be used, as shown in Figure 4.

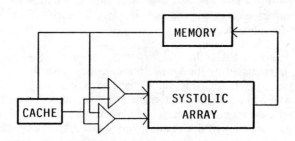

Figure 4.   Use of cache to buffer lines from the input image.

PSC Demonstration System

Figure 5 depicts the current PSC demonstration system built around a SUN workstation. The system includes a PSC array board capable of holding 25 or more PSCs. As of May

215

1984, this board hosts a one-dimensional systolic array of nine PSCs, performing 2-D convolutions using 3×3 kernels on a video image of 512×512 8-bit pixels. Several 3×3 Gaussian and Laplacian kernels have been implemented for the demonstration. In the demonstration, 512×512×8 displays are processed at the rate of one display every 1.8 seconds. Using the "scalability" of the systolic array design as described in the previous section, the demonstration system can run at this speed *regardless* of the kernel size (assuming of course that there are as many PSCs in the PSC array board as the kernel size). For instance, with 25 PSCs the demonstration system can perform 2-D convolutions using 5×5 kernels still in the same 1.8 seconds. Indeed, it is our plan to do such a demonstration, as soon as enough working chips are acquired.

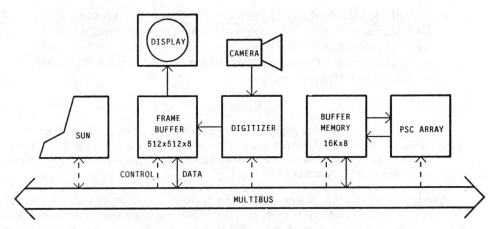

Figure 5. PSC demonstration system.

The host system for the demonstration is the SUN workstation that controls the PSC array board through its MULTIBUS interface. A buffer memory board buffers data for the PSC array board. Matrox graphics boards are used to perform the frame buffering and video A-to-D, D-to-A functions. The limited bandwidth of the Matrox DMA imposes one of the speed limits for the demonstration system. Besides the PSCs themselves, the PSC array board also contains a finite state controller, clock drivers, microcode loading circuitry, and address generation circuitry for the buffer memory board.

The demonstration system operates as follows:

1. The SUN's 68010 processor initializes the PSC array by writing to its control registers, which are mapped into the Multibus address space.

2. The 68010 loads the microcode of each PSC individually.

3. The 68010 initializes the Matrox graphics boards.

4. The Matrox VAF-512 board grabs one frame of video data from the camera, and loads it into the frame buffer.

5. The 68010 initiates a DMA transfer of five video lines from the frame buffer to the PSC buffer memory.

6. When above transfer is complete, the 68010 starts the PSC array, at a buffer address which it has supplied.

7. The PSC array reads in five lines of data, and produces three lines of output, writing it into the buffer. The 68010, meanwhile, initiates a DMA transfer of the next three lines of data into the buffer, which takes place in parallel with the PSC processing.

8. The 68010 periodically checks the status of the PSC array, and, upon sensing the DONE flag, restarts the array at the next buffer location. It then begins a DMA transfer of output data into the Matrox frame buffer. When the output transfer is completed, the 68010 initiates another DMA of input data to the PSC buffer, and waits for the PSC DONE flag.

Hindsights

The design of the PSC is a moderately large project by university standards. As of August 1984, the total effort has been about 4 man-years, with the following rough breakdown: architecture (1), logic and circuit design (.5), layout (.7), testing (.5), demonstration system (.8), and tool development (.5).

As often occurs in large, experimental system projects, the nature and demands of the PSC project did not become clear to us until the project was more than half-way through. We are pleased that the chip works and a modest demonstration system is running. At the same time, it is disappointing that up to now we have not been able to run large demonstrations, or to do experiments with many applications such as the decoder implementation for Reed-Solomon error-correction codes.

The PSC has not been applied on a large scale for two basic reasons. First, fabrication yield of the PSC has been low and has varied substantially over different MOSIS runs, and as a result it is difficult to predict when we will have a large number of working chips. Second, as the first of its kind, the PSC has no sophisticated software and interface support. This has prevented substantial applications from being brought up on the PSC at this time.

There were also many problems encountered in building, testing, and demonstrating the PSC. Some of these problems are inherent to the fact that the PSC project is experimental, and that the resources available to the project have been severely limited. (As far as we can tell, the cost of the PSC project is no more than one hundredth of the development cost of a typical commercial microprocessor!) However, there are a number of things that we would definitely do differently next time. In the following we discuss some of these hindsights under three categories: architecture, design, and management.

Architecture

The PSC architecture seems to be very well-suited to the implementation of a wide variety of systolic algorithms. The 64 instruction words available accommodate most straightforward computations, and suffice even for such complex cell computations as those found in the systolic Reed-Solomon decoder. The processor's control structure imposes very little overhead relative to the arithmetic heart of the computation; instruction fetch occurs in parallel with execution. Concurrency and parallel I/O result in much smaller instruction counts than for conventional microprocessors. Finally, the incorporation of control and communication circuitry onto the same chip as the arithmetic units achieves a large savings in chip count over systems built with standard parts. A PSC equivalent built with LSI arithmetic, memory and control and with TTL latches and multiplexers would require on the order of one hundred chips, and a similarly constructed single-purpose cell for even a slightly complicated algorithm would require a dozen chips.

However, the current PSC architecture is not completely optimized, partially for reasons of simplicity and flexibility that were considered important in view of the experimental nature of the PSC project. As our insights into the nature of systolic computation have increased, we have found several improvements that can be made. Some of the improvements described below have already been incorporated into the design of the CMU Warp processor, now under development.[9, 10]

For simple computations, the PSC's arithmetic, internal communication and external communication capacities are fairly well balanced. For more complex algorithms, the most common limiting factor in program performance is the number of buses. Adding more buses would be quite expensive in area needed for routing and for code storage and distribution; for a general-purpose part, this expense would probably not be justified. Other possibilities would be to use some specialized buses, to allow a single bus to be broken into independent parts, or to multiplex the use of the buses within a machine cycle. One other option, which seems to be very useful in the frequent case where a value needs to be delayed as it enters or leaves a cell, would be to put small programmable delays on the chip's input or output ports.

Another limitation is in the bandwidth of the register file. Only one word can be read or written in a cycle, making the storage and retrieval of intermediate results time-consuming compared to computation. Possible improvements include the use of multiported registers.

For large filtering problems where large numbers of terms may be accumulated at each cell, the multiplier-accumulator needs a high-precision accumulator. Preferably, the width of the accumulator should be at least 24 bits.

One way to reduce the cost of a PSC-based system would be to reduce the chip's complement of I/O pins, 54 of which are dedicated to the data ports. The PSC's use of three input and three output ports is due mainly to simplicity considerations: almost all systolic algorithms' data flows can be implemented in a straightforward way with a minimum of control. Since all six ports are needed simultaneously only for rather simple algorithms where communication dominates computation by a large factor, it may be possible to reduce the pincount of the chip without greatly reducing its overall performance. This could be achieved by multiplexing bidirectional ports (perhaps four), at a modest cost in control complexity.

Another area where the cost of the PSC might be reduced is in microcode space. Again for reasons of simplicity and flexibility, no attempt was made to squeeze the microinstruction size by limiting the number and kind of operations the PSC's parallel functional units could perform. While this is a useful property for an experimental system, it would be advantageous for production chips to sacrifice some flexibility for higher yield (due to smaller size) or more words of data or instruction memory.

Design

As mentioned earlier, the PSC project was the first major chip design effort at CMU. The actual process of bringing the PSC from its initial architectural concept to the current demonstration system has been a great learning experience. Part of the teaching was done by some serious technical difficulties, which were mostly related to chip operating tolerances (clock waveform and supply voltage sensitivity), yield and programming.

Electrical design is probably the weakest point in the PSC design. The memories, especially, have been less than robust over voltage, temperature, and clock waveforms. The yield problem has been mostly due to failures in dynamic RAM.

A related problem is the complex timing scheme used, which necessitated many off-chip clock signals, making speed testing difficult. The complexity of the memory timing scheme resulted in several patches being applied; it would have been better to clean it up.

A 700 ns cycle time has been observed for some PSC prototypes, but many of the prototypes have been found to run around 1.2 μs. One reason that the chip is not as fast as possible is that performance tuning of the layout was never done, for example, for the multiplier-accumulator circuit; speed was not one of the project's primary goals, and timing analysis tools like Berkeley's Crystal[11] and Stanford's TV[12] were not available. Another contributor has been a drift in MOSIS circuit parameters. The chip was designed under the assumption that diffusion resistance was 10 ohms/square, as in Mead/Conway (and as assumed, by default, by Crystal). A number of MOSIS runs have had resistances of 11 or 12 ohms/square. Under this assumption, the microcode bits have an estimated maximum delay of 50 ns. Recent MOSIS runs have had a diffusion resistance of 40 ohms/square, increasing that figure to 200 ns.

Since the available simulators capable of handling large numbers of transistors were not capable of handling the memory and some other features of the chip, full-chip layout simulation was never done. All of the pieces were tested and/or simulated, but test results in some cases were not available until after the entire chip had been sent off. At the time the chip was assembled, the only means available of checking connectivity of the parts was manual inspection of a 60 page condensed wirelist. This process caught one or two bugs, but one bug slipped by. Chip testing, though, was fairly straightforward: except for the sequencer, which could be tested only if the memory worked, everything on the chip was accessible over the buses.

A lack of good documentation is one reason the design was difficult to change. Another, bigger reason was the difficulty of routing. The PSC contains a lot of square microns of wire, painfully drawn by hand. Turnaround time was another factor discouraging major changes.

The chip had only three layout bugs, which were corrected early on; one was in the memory, one in one set of ALU registers, and one in the multiplier condition code (hard to notice, since the condition codes are very obscure). There was one logic bug (discounting the memory's problems, which were mostly electrical); it related to the timing of the sequencer's stack, was generated by a change in memory timing and was caught very late (summer '83) because of the dependence of circuit testing on the memory and because of a lapse in testing effort between March and June.

Many of these difficulties could be avoided next time by using appropriate design methodologies, conservative design styles, and new design tools. Some of the improvements that we can make seem to be rather "common sense" and obvious now. These include designing for worst-case technology and avoiding complex timing schemes. Given our multi-source fabrication through MOSIS, we must be prepared to deal with a relatively large process variation. Specifically, clever but risky circuit design should not be used. Mead and Conway simplified design rules exist, in part, for this purpose. The example of diffusion resistance mentioned earlier shows that pessimistic assumptions are important in the electrical domain, as well. Complex timing should be avoided because it will make the design difficult to understand, hard to change, and hard to deal with when testing and interfacing to the chip. The gain in performance through complex timing is hardly worth its cost.

As mentioned earlier, new timing aids like Crystal and TV will help remove critical timing paths. Using modern design workstations, such as Daisy and Mentor, schematics of the chip can be fully simulated and documented at the logic level with reasonable effort. Also, these workstations can generate netlists to be compared with those extracted from the layout, using the CMU wirelist comparison program, Gemini.[13]

A careful floorplan can also make a lot of difference in the quality of the design. Since useful global routers are still not available, a good floorplan can make the routing problem much easier. The floorplan should be drafted in the very beginning of the design, and constantly updated during the design as detailed layout information becomes available. From the floorplan one can tell if a certain design optimization is worthwhile. Also, the availability of the updated floorplan can substantially help the communication between a team of designers.

Management

Building a prototype research chip in a university environment is very different from the same task in an industrial environment, in the sense that we are severely limited by resources. It would be unreasonable to expect that universities have the same level of design support and engineering skill as semiconductor industry. Graduate students should be involved but shouldn't be expected to grow old in the course of a project. Probably one of the biggest lessons we have learned from the PSC project is a true appreciation of this limitation in resources.

But prototype chips, built by universities or not, must work in a system, otherwise there would be very little value in the prototyping. Therefore it is important that we do only those designs which are within the power of the available people and tools. Designs that industry does well and that require great skill and experience should be avoided by universities. Fancy dynamic RAM is an example.

An overall plan for simulation, verification, documentation, testing, and demonstration should be developed at the very beginning of a project. We must see to it that there is a very good chance that the chip will work at reasonable speed in a system for its first silicon, and that system demonstrations can be brought up without an excessive amount of effort. A

220

related issue is the yield problem. We must learn (at last!) to trade architectural features for reduced die sizes in order to increase yield. These steps are necessary conditions for a large scale chip to be built successfully and smoothly. It is useful to remind ourselves that university prototype chips are not exempt from that.

The lack of sufficient system support mentioned earlier is a common problem for any prototype chips. One way to deal with it is to stage research programs so that software and interface support can be developed first on systems built with off-the-shelf components. Ideally, prototype custom chips should be built only after system support has already been developed.

Conclusions

Despite the problems discussed above, the PSC project produced a number of positive results. The chip works, albeit not as robustly as we would like. From the architectural point of view, the project demonstrated the "scalability" of systolic array design in the demonstration system, proved the feasibility of having a programmable "building-block" chip for the implementation of systolic algorithms and, through setting a concrete benchmark on which to base improvements, set the stage and provided initial ideas for further work. A natural step to follow is the development of an industrial version of the PSC. Several companies have expressed their interests in this. In theory, companies who produce PSC-like chips should be able to sell hundreds of copies of the chip to each customer, to form large systolic-like arrays!

The PSC project did not contribute to the low-level chip design knowledge of the world at large, but we learned a lot of things locally about chip design, both personally and in terms of the VLSI community at CMU. This includes not only the lessons mentioned above, but also the use of new tools and methods. The PSC experience has had profound impacts on the ways in which how some new CMU chips are being designed, as suggested in the preceding section.

The PSC is one of the first major chips made through MOSIS to have been integrated into a system. Work on the PSC also helped the MOSIS community gain experience in packaging, testing, and medium-volume production.

Acknowledgments

The PSC is a result of a team effort; its architecture and design have been reported in separate papers,[2, 3, 4] on which some of the material of this paper is based. F. H. Hsu wrote the PSC image processing code for the demonstration system. Some chip testing software was developed by Monica Lam. The PSC research was supported in part by the Defense Advanced Research Projects Agency (DoD), ARPA Order No. 3597, monitored by the Air Force Avionics Laboratory under Contract F33615-81-K-1539. The fabrication of the PSC has been done through MOSIS, the DARPA silicon broker.[14]

References

1. Kung, H.T., "Why Systolic Architectures?," *Computer Magazine*, Vol. 15, No. 1, January 1982, pp. 37-46.

2. Fisher, A.L., Kung, H.T., Monier, L.M. and Dohi, Y., "The Architecture of a

Programmable Systolic Chip," *Journal of VLSI and Computer Systems,* Vol. 1, No. 2, 1984, pp. 153-169, An earlier version appears in *Conference Proceedings of the 10th Annual Symposium on Computer Architecture,* Stockholm, Sweden, June 1983, pp. 48-53.

3. Fisher, A.L., Kung, H.T., Monier, L.M., Walker, H. and Dohi, Y., "Design of the PSC: A Programmable Systolic Chip," *Proceedings of the Third Caltech Conference on Very Large Scale Integration,* Bryant, R., ed., Computer Science Press, Inc., California Institute of Technology, March 1983, pp. 287-302.

4. Walker, H., "The Control Store and Register File Design of the Programmable Systolic Chip," Tech. report CMU-CS-83-133, Carnegie-Mellon University, Computer Science Department, May 1983.

5. MacWilliams, F.J. and Sloane, N.J.A., *The Theory of Error-Correcting Codes,* North-Holland, Amsterdam, Holland, 1977.

6. Peterson, W.W. and Weldon, E.J., Jr., *Error-Correcting Codes,* MIT Press, Cambridge, Massachusetts, 1972.

7. Kung, H.T., Ruane, L.M., and Yen, D.W.L., "Two-Level Pipelined Systolic Array for Multidimensional Convolution," *Image and Vision Computing,* Vol. 1, No. 1, February 1983, pp. 30-36, An improved version appears as a CMU Computer Science Department technical report, November 1982.

8. Kung, H.T. and Picard, R.L., "One-Dimensional Systolic Arrays for Multidimensional Convolution and Resampling," *VLSI for Pattern Recognition and Image Processing,* Fu, King-sun, ed., Spring-Verlag, 1984, pp. 9-24, A preliminary version, "Hardware Pipelines for Multi-Dimensional Convolution and Resampling," appears in *Proceedings of the 1981 IEEE Computer Society Workshop on Computer Architecture for Pattern Analysis and Image Database Management,* Hot Springs, Virginia, November 1981, pp. 237-278.

9. Kung, H.T., "Systolic Algorithms for the CMU Warp Processor," *Proceedings of the Seventh International Conference on Pattern Recognition,* International Association for Pattern Recognition, 1984, pp. 570-577.

10. Kung, H.T. and Menzilcioglu, O., "Warp: A Programmable Systolic Array Processor," *Proceedings of SPIE Symposium, Vol. 495, Real-Time Signal Processing VII,* Society of Photo-Optical Instrumentation Engineers, August 1984.

11. Ousterhout, J. K., "Crystal: A Timing Analyzer for nMOS VLSI Circuits," *Proceedings of the Third Caltech Conference on Very Large Scale Integration,* Bryant, R., ed., Computer Science Press, Inc., California Institute of Technology, March 1983, pp. 57-70.

12. Jouppi, N.P., "TV: An nMOS Timing Analyzer," *Proceedings of the Third Caltech Conference on Very Large Scale Integration,* Bryant, R., ed., Computer Science Press, Inc., California Institute of Technology, March 1983, pp. 71-86.

13. Ebeling, C. and Zajicek, O., "Validating VLSI Circuit Layout by Wirelist Comparison," *Proceedings of 1983 IEEE International Conference on Computer-Aided Design,* IEEE, September 1983, pp. 172-173.

14. Lewicki, G., Cohen, D., Losleben, P. and Trotter, D., "MOSIS: Present and Future," *Proceedings of Conference on Advanced Research in VLSI,* Penfield, P. Jr., ed., Artech House, Inc., Massachusetts Institute of Technology, Dedham, Massachusetts, January 1984, pp. 124-128.

PART IV:
TESTING OF VLSI CHIPS

MICROARCHITECTURE OF THE MC 68000 AND EVALUATION OF A SELF CHECKING VERSION

P.MARCHAL, M.NICOLAIDIS, B.COURTOIS

IMAG/TIM3
46, avenue Félix Viallet
38031 GRENOBLE CEDEX
Tel. (76) 47 98 55

Abstract

The microarchitecture of the MC 68000 microprocessor is examined, including technological and system characteristics, resources, instructions, signals exceptions and test mode. The internal architecture is detailed: data processing section and control section . Next basic rules for the design of self-checking NMOS circuits are briefly given. These rules are based on fault hypotheses including for example transistors s-open/s-on, shorts between aluminum lines, etc... The application of these rules to the design of a self-checking version of the MC 68000 is not detailed, but an evaluation of what should be such a redesign is given. The silicon overhead is detailed.

The work reported hereafter has been realized by the Computer Architecture Group belonging to the TIM3-IMAG-INPG Laboratory. It has been a part of a larger study aimed at understanding the microarchitecture and fine functioning of the MC 68000, in order to evaluate that microprocessor for specific applications, to study testing of that microprocessor and finally to evaluate what could be a self-checking version of that microprocessor. The only documents available were data sheet and papers published in the litterature. The instruments used to analyze the circuit are optical and electronic microscopes (note that Computer Architecture Group has an extensive experience in reverse engineering, for several years). Numerous papers have been published and test programs have been written. In the following, only the micro-architecture and the evaluation of a self-checking version are discussed.

The summary of this paper is as follows :

Summary

PART A - MC 68000 Presentation
PART B - MC 68000 Microarchitecture
PART C - Design of NMOS Self-checking Circuits
PART D - Evaluation of a Self-checking MC 68000
PART E - Conclusions.

227

PART A - MC 68000 PRESENTATION

1 - Introduction

2 - General Characteristics
        1 - Technological Characteristics
        2 - System Characteristics

3 - Resources
        1 - Programmer Accessible Registers
        2 - Hidden registers

4 - Instructions
        1 - Instruction set
        2 - Addressing modes

5 - Signals and Buses
        1 - Address Bus
        2 - Data Bus
        3 - Control Bus
              3.1 Asynchronous Bus Control
              3.2 Bus Grant Control
              3.3 Interrupt Control
              3.4 System Control
              3.5 MC 6800 Family Control
              3.6 Processor State

6 - Exceptions
        1 - Internal checking exception
        2 - External checking exception
        3 - Exception processing

7 - Test Mode

1 INTRODUCTION

Since 1980, microprocessor technology has been enterring a new
and especially challenging era. The limiting factor is no more
the silicon area or the technology constraints but creativeness
and imagination of designers.
This chapter introduces one of those 16 bits monochip, the MC
68000 microprocessor designed and manufactured by MOTOROLA inc.
The MC 68000 design was completed in late 1979 at the end of
a very intensive project by a MOTOROLA designers group. The
design time has been evaluated, by MOTOROLA [FRA 80] to be 100
man-months, the layout time has been evaluated to be 70 man-
months, while the elapsed time to first silicon has been 30
months.
Although the MC 68000 is representing the same amount of design

and layout of effort as the Z8000 - another 16 bits monochip
microprocessor designed by Zilog- the Z8000 debugging time was
three times longer (18 months) than the MC 68000 one (6 months).
At the time, this machine represented a considerable advance
over the other integrated CPU available.

The MC 68000 marked a very large clear change in the internal
architecture of integrated processors.

Its internal organisation is fairly representative of those
I.C. wich implement mini and medium range computers.

Functionally its organisation is fairly classical. In particular,
the micro-programmed control uses microaddress bit -replacement
techniques, which were used in the IBM 360 series of computers
as early as 1964.

Moreover, it should be noted that, because of its internal architecture optimisation needed for both high performance and
acceptable silicon area, the MC 68000 is tied to its instruction
set. It cannot be considered as a 'microprogrammable' machine
in the sense that most people use the word. This loss of gene-
ralites was the price that had to be paid in order to reduce
the microprogram area by a factor of approximately 4.

Topological problems have had little influence on the functional
organisation of the control part and especially in the central
parametrisation area where regularity is very much reduced.

The most elegant part is without any doubt the operative part
which provides an almost perfect solution to functional, electrical and topological problems.

The MC 68000 die photograph of picture 1 depicts the regular
internal organisation.

2 - GENERAL CHARACTERISTICS

2 - 1 Technological characteristics

- launched in mid 1980
- HMOS 1 technology with depletion loads
- 3.5 microns polysilicon gates (new version HMOS2 due
  to appear in 2 microns technology)
- four transistor-types are used :
  . enhancement mode
  . natural mode
  . lightly implanted depletion mode
  . heavily implanted depletion mode
- chip area 68880 mils ($44,43$ mm^2)
- complexity 68000 "potential" transistors
- regularity ratios (layout MOS/designed MOS):
  . PLA - ROM : 50/1
  . register 500/1
  . random logic : 5/1
  . weighted average : 74/1

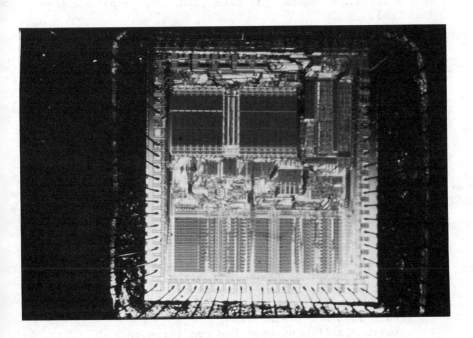

Figure 1 : MC 68000 die photograph

- supply : 5 volts
- automatic substrate bias voltage generation : -2.5 Volts
- static and dynamic circuitry
- external clock frequency :
    . MC 68000 F4 - 2 to 4 MHZ
    . MC 68000 F6 - 2 to 6 MHZ
    . MC 68000 F8 - 2 to 8 MHZ
    . MC 68000 FA - 2 to 10 MHZ

2 - 2 System characteristics

- 64 pins package,
- instruction set containing 56 basic instructions,
- 14 addressing modes,
- 1, 8, 16, 32 data length operations,
- 1 to 5 16-bits word instructions format,
- internal clock with 2 sequences levels :
    . basic non-overlapping phase signals $\emptyset 1$ and $\emptyset 2$ running
      at the external clock frequency.
    . 4 successive instant signals T4, T1, T2, T3 which
      span two $\emptyset 1$ and $\emptyset 2$ periods,
- instruction execution between 2 and 88 instants signal
  cycles,
- non multiplexed buses : 16 bits data bus, 23 bits address
  bus (24),
- internal generation of signals for controlling asynchronous
  external transfers,
- generation, on external request, of fully MC 6800 family
  compatible signals,
- vectored interrupt mechanisms with 7 priority levels,
- test mode (4% of the chip area).

3 - RESOURCES

3 - 1 Programmer accessible registers

As shown in figure 2, the programmer accessible registers are
the following :
- 8 32-bit data registers D [0=7],
- 7 32-bit address registers A [0=6],
- 2 32-bit stack-pointers :
    . +1 user stack-pointer A [7],
    . +1 superviser stack-pointer $_A$' [7],
- 1 32-bit program counter (only 24-bit are actually used
  for address generation),
- 1 16-bit status register PSW :
    . least significant byte for condition code CC,
    . most significant byte for system condition CS.

In fact, as shown in figure 3, only 5 flip-flops are available in both bytes.

3 - 2 Hidden registers

Some more internal registers are used by the machine itself but are unknown by the programmer. These are the following :
- 2   32-bit temporary registers:
    . 1 data temporary register
    . 1 address temporary register,
- 1   32-bit address output buffer-register (outputs only 23 bits),
- 2   16-bit data buffers
    . 1 input data buffer
    . 1 output data buffer,
- 1   16-bit ALU-buffer,
- 7 more 16-bit registers each of them embedded in special operators.

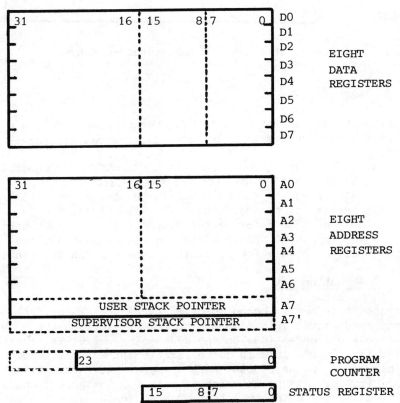

figure 2 : MC 68000 programmer accessible registers

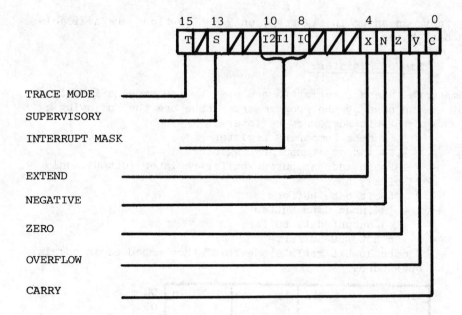

Figure 3 : internal structure of the status register

4 THE INSTRUCTIONS

The instructions consist of a 16-bit operation-word followed
by 0 to 4 words for operands or addresses. We can distinguish
between :
- 4 bit op-code : for 2 operands instructions or conditional
  instructions (4 more bits for the condition).
- 10 bit op-code : for 1 operand instructions, the operand
  is precise by the "effective address" field,
- 12 bit op-code : for vectored addressing instructions,
- 13 bit op-code : for register addressed instructions,
- 16 bit op-code : for implicit operands instructions.

4 - 1 Instruction Set

The instruction set is composed of 56 types of instructions
as mentioned above. This seems to be few but in fact the number of possible codes (due to the addressing capabilities) is
greater than 45 000,this is quite large compared to the 192
possible codes on the M 6800. Further more, the instruction
set is high-level language oriented with for instance : double
condition looping primitive, test-and-Set instruction, large
possibilities in stack-operations (for a more detailed description,

refer to the manufacturer's data sheets).

4 - 2 Addressing Modes

The MC 68000 allows 14 different addressing modes. These addressing modes may be precised inside the operation word or may necessitate one or two extension words. The different addressing modes are the following :
- direct register addressing (2) :
  - . Data register direct,   op = Dn
  - . Address register direct ;   op = An
- indirect register addressing (7) :
  - . indirect address register,   op = (An)
  - . postincremented indirect address register,   op = (An)+
  - . predecremented indirect address register,   op = -(An)
  - . indirect address register relative,   op = d(An)
  - . program counter relative,   op = d(Pc)
  - . indexed indirect address register,   op = d(An,Pi)
  - . program counter indexed ;   op = d(Pc,Pi)
- absolute addressing (2) :
  - . short absolute,   op = xxxx
  - . long absolute ;   op = xxxxxxx
- immediate addressing (2) :
  - . immediate implicite,   op = #xxx
  - . immediate explicite ;   op = #xxx
- implicite addressing   op = CCR,USP,

5 - SIGNALS AND BUSES

As shown in figure 4, the MC 68000 is housed in a 64-pin package that allows separate (non-multiplexed) data and address buses.

5 - 1 Address Bus

The address bus is composed of 23 pins A[1=23]. Although all address evaluations are performed over 32 bits only the 24 least signifiant bits are used to generate an external address (bit 0 generates $\overline{UDS}/\overline{LSD}$). This configuration allows in conjunction with $\overline{UDS}/\overline{LSD}$ a memory addressing range of 16 Mega-bytes.

5 - 2 Data Bus

The data bus is composed of 16 pins D [0=15]. It is a three-state bidirectional bus. During interrupt cycle, lines D0 through D7 are driven by the external device to provide the interruption vector number.

234

5 - 3 Control Bus

The control bus is 20-bit wide, it is composed of three sub-control buses.

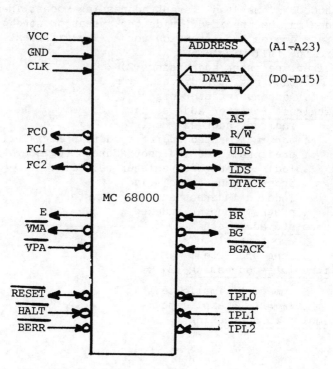

Figure 4 : MC 68000 pins identification.

5 - 3.1 Asynchronous bus control.

This sub-control is 5-bit wide, it controls asynchronous input/
output transfers with memory and external devices. These 5 lines
are the following :
- address strobe ($\overline{AS}$),
- read/write ($R/\overline{W}$),
- upper data strobe ($\overline{UDS}$) and lower data strobe ($\overline{LDS}$) both
  generated from bit 0 of the address bus,
- data transfer acknowledgement ($\overline{DTACK}$).

The asynchronous bus exchange uses the whole possibility of the
MC 68000. For an asynchronous protocol, two transfers may be
distinguished whether or not the internal clock is stopped. In fact,
it depends whether or not the transfer result (read or write) is
necessary during the microinstruction where the transfer is
started.

In the first case, a special line stops the internal timing (the four instant signals). There is only the two basic phase Ø1, Ø2 running at the external clock frequency until the transfer acknowledgement signal (DTACK) is asserted.
This protocol enables the interconnection of memory running at any speed. The figure 5 depicts such an example of transfer including a waiting phase.
Upon receiving the DTACK signal, the processor reruns the instant signals, the execution of the microinstruction is then restarted. After a maximum delay, if the DTACK signal does not come, the MC 68000 enters a Bus Error Exception processing.

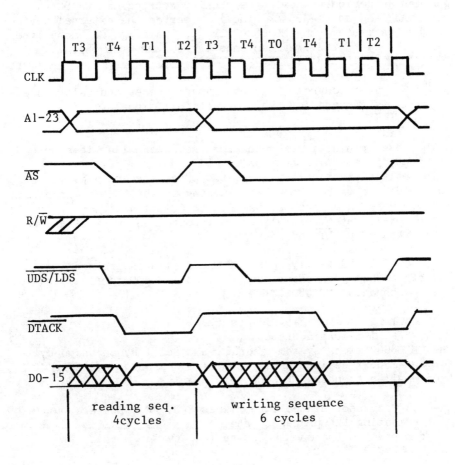

Figure 5 : MC 68000 asynchronous bus transfer

5 - 3.2 Bus Grant control. This sub-part manages the bus grant.
It is composed of three signals :
- bus request, ($\overline{BR}$),
- bus Grant ($\overline{BG}$),
- bus Grant Acknowledgement ($\overline{BGACK}$).
These three signals are driving the bus grant automata. This
technic enables external devices to request, to get and to
recognize bus grant. So, in order to get the bus the requesting
device drives the $\overline{BR}$ line high, and then it waits for the acknowledgement. Upon receiving the request signal on its $\overline{BR}$ line ,
the bus arbitration automata of the MC 68000 asserts the bus
grant lines while ending the started input/output cycle. The
bus master device acknowledges then by setting up its $\overline{BGACK}$
line. As long as the $\overline{BGACK}$ line is asserted the external device
is the bus master, the MC 68000 is just waiting. One more line
can be added to the previous ones : the $\overline{BERR}$ line which is
activated during Bus Error occurrences.

5 - 4.3 System control. This bus manages reset and halt processing. It is composed of two bidirectional signals :
- $\overline{reset}$,
- $\overline{halt}$.
. The reset pin activation enables the external devices to
be reseted (reset instruction).
. The halt pin activation stops the processor and puts it into
halt mode : a microinstruction jumping to itself. The processor
keeps on working as soon as the pin is desactivated.
. The total system initialisation is obtained by simultaneous
activation of $\overline{RESET}$ and $\overline{HALT}$ pins.

5 - 3.5 MC 68000 family control. These signals are used to
interface the synchronous devices of the M6800 family with
the asynchronous MC 68000 bus :
- enable E,
- valid peripheral address ($\overline{VPA}$),
- valid memory address ($\overline{VMA}$).
The MC 68000 clock system generates a special synchronous cycle
for the M 68xx devices. This special synchronous cycle is 10
time smaller than the MC 68000 cycle.

5 - 3.6 Processor state. Three lines FC0, FC1, FC2 are output
pins indicating the processor state.
FC2 pin precise the machine state (supervisor/User).

FC0	FC1	
0	1	data area
1	0	program area
1	1	interruption acknowledgement

Five more lines are to be added :
- the power lines (2),

- the ground lines (2),
- the clock line (1).

6 - EXCEPTIONS

The exception mechanism is composed of two parts:
- either internal checking exception mechanism,
- either external checking exception mechanism.
A 1024-byte table giving 256 program counters starting addresses
are necessary to ensure all exception processing.

6 - 1 Internal Checking Exception

Two cases can be derived from those internal checking exceptions :
- either by the execution of an instruction such as :
  . trap
  . trapv
  . check
The exception is then generated by the instruction itself
(corresponding to a normal behaviour).
- or by an abnormal execution such as :
  . invalid op-code detection,
  . privilege violation,
  . trace mode,
  . bus error,
  . address error,
  . divide by zero.

6 - 2 External checking exception

Such exception are checked by external devices. Two cases can
be derived from those external checking exceptions:
- reset and halt pins activation ,
- interrupt pins activation.

6 - 3 Exception processing

Exception processing occurs in four steps :
. first step : internal copy of status register which is then
              set to exception processing (S-bit set, T-bit
              reset),
. 2nd step   : exception vector number determination to get
              the index in the exception table,
. 3rd step   : status register, current program counter stacked,
              some more information are stacked depending
              on the type of exception,
. 4th step   : processor resumes instruction execution by fet-
              ching the first instruction given by the address

of the exception vector.

7 - TEST MODE

The MC 68000 has a special mode for simplifying the test of
its control part (the test of the data processing section is
much more easier due to its higher observability and controla-
bility).
This mode is enabled by taking the $\overline{VPA}$ pin to approximately
8 volts. The special hardware activated for this mode occupies
approximately 4% of the whole chip area and involves around
250 MOS transistors.
In that mode the control part is directly exercised after input
pins reconfiguration. The microinstruction contents is readable
by third on address pins.
The behaviour of some of the internal units is modified as
follows :

Data processing part :
- data pin output drivers are disabled,
- data input gates from the pin are permanently enabled
  for operation word inputting,
- the input/output byte duplication mechanism is inhibited,
- the address pin output drivers are disabled,
- the data processing ALU is disabled.

Control part :
- loading of the parametrisation PLA is inhibited,
- each third of the action part of a microinstruction can
  be connected to the parametrisation PLA lines (23 lines
  of the 32 are used, the others are forced to zero).
  The appropriate third is chosen by the value of pins
  IPL1, IPL2,
- the 23 lines of the parametrisation PLA used to carry
  the third of microinstruction are connected to the
  address pins.
- the outputs of the conditional branching PLA are replaced
  by the value of $\overline{BR}$ and $\overline{BGACK}$ pins,
- the selection of the decoding PLAs (A1, A2, A3) is made
  by the value inputed on pins $\overline{DTACK}$ and $\overline{IPL\ 0}$.

DTACK	IPLO	MAR3	MAR2	Selected PLA
0	1	X	X	PLA A1
1	0	X	X	PLA A2
1	1	X	X	PLA A3
0	0	0	1	PLA A1
0	0	1	0	PLA A2
0	0	1	1	PLA A3
0	0	0	0	next address field

Figure 6 : decoding PLA selection in test mode.

PART B - MC 68000 MICROARCHITECTURE

1 - Data Processing Section
    1 - Introduction
    2 - High-order Address Processing Sub-part
    3 - Low-order Address Processing Sub-part
    4 - Data Processing Sub-part
    5 - General Features of all three Processing Sub-parts
      5.1 Carry Look-ahead
      5.2 Bus System
      5.3 Clock System

2 - Control Section
    1 - Introduction
    2 - Microprogram ROM
    3 - Decoding PLA A1,A2 and A3
    4 - Exception PLA A0
    5 - Conditional branching PLA
    6 - Parametrisation PLA
    7 - Sequencing of the control section
    8 - status register
    9 - Pre-Fetch Mechanism

1 - DATA PROCESSING SECTION

1 - 1 Introduction

The MC 68000 data processing section has a structure composed
of perfectly rectangular slices with metal buses along the
length and polysilicon control lines across the width. The
very rectangular layout of this part is obtained by assembling
a relatively small number of cells on a double bus structure
which crosses the entire data processing part.
The data processing section is managed by 196 control lines, it
returns to the control section 23 state signals and a 16-bit
bidirectional bus which enables to store or load the status
register. We can notice that there are also two state signals
generated by the data processing section which are not connected
in the control part. Functionally, this data processing part
is composed of two processing sub-part :
    - a 16-bit data processing sub-part,
    - a 32-bit address processing sub-part.
In order to achieve a more regular design, the 32-bit address
processing sub-part has been split into two 16-bit sub-parts
which process the lower and the upper halves of the addresses.
These three processing sub-parts use the same double-bus system
interrupted by two sets of switches which allow data transfers
between two parts or independent processing in each part.
For convience but also for a time consumming point of view,
the upper half address processing sub-part also contains the
upper half of 32-bit data registers.

1 - 2 High Order Address Processing Sub-part

The figure 7 shows the functional scheme of the high order
address processing sub-part.
This processing sub-part, referred to as HAPP, contains :
    - the upper half of the following external architecture
      registers :
        . address registers AH [0:6],
        . data registers DH [0:7],
        . user stack pointer AH [7],
        . supervisor stack pointer AH [7'],
        . program counter PCH ;
    - the upper half to two working registers :
        . address temporary results ATH,
        . data temporary results DHT ;
    - the upper half to the address output buffer AOBH connec-
      ted to the address pads A <16 : 23> ;
    - the upper 16 bits of the 32-bit arithmetic unit which
      mainly performs address calculations.

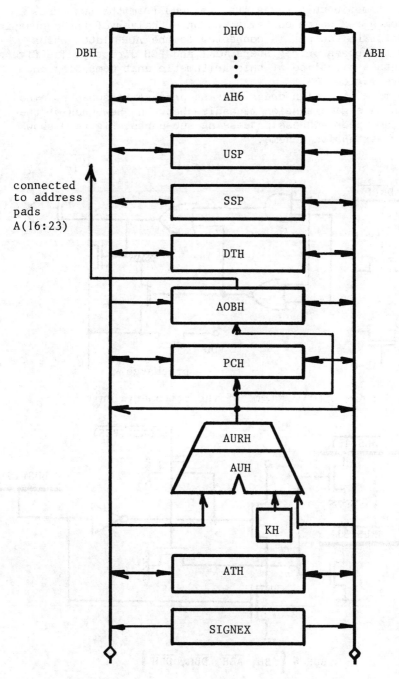

Figure 7 : Functional scheme of the high order address processing sub-part.

242

This arithmetic unit takes its operands from the two buses
and from a ROM which contains the upper halves of the addressing
constants. Its output is connected to the AURH latch which
drives the buses or registers AOBH and PCH directly. The figure
8 depicts a bit-slice of this arithmetic unit computing on
negative logic.
Furthermore, the high order address processing sub-part also
contains a sign extension mechanism between the buses of the
lower and upper address processing sub-parts. Figure 9 shows
the sign-extension scheme.

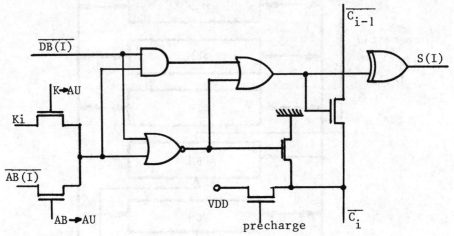

Figure 8 : Electrical scheme of the arithmetic unit.

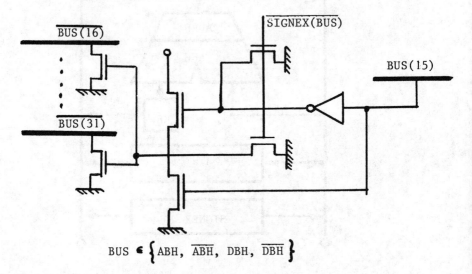

$$BUS \in \left\{ ABH, \overline{ABH}, DBH, \overline{DBH} \right\}$$

Figure 9 : Electrical scheme of the sign-extension operator.

1 - 3 Low Order Address Processing Sub-part

The figure 10 shows the functional scheme of the low order
address processing sub-part. This processing sub-part, referred
to as LAPPP contains :
- the lower half of the following external architecture
  registers :
  . address register AL [0:6],
  . user stack pointer AL [7],
  . supervisor stack pointer AL [7'],
  . program counter PCL ;
- the lower half of the working register for address temporary results ATL ;
- the lower half of the address output buffer AOBL connected
  to the address pads A<1 = 15> (bit AOBL(0) is directly
  transmitted to the control part for address error exception
  processing and is also used to drive pins $\overline{UDS}$ and $\overline{LSD}$) ;
- the 16-bit register FTU, for bidirectional interfacing
  with the bus BC which span the control part to the data
  processing part ;
- the lower 16-bit of the arithmetic unit AUL. The lower
  half of the arithmetic unit and the higher half are the
  identical, the lower half also takes its operands from
  the two buses and from a ROM which contains the lower
  halves of the addressing constants 0, +1, +2, +4, -4,
  -2, -1. Its output is also connected to a AURL latch
  which drives the buses or the registers AOBL and PCL
  directly ;
- the priority encoder, referred to as PREN operator
  connected to the A bus stores the content of this bus
  and detects the existence of at least one bit at '1'.
  It encodes, for the control part, the index of the first
  bit at '1' starting from the least significant bit,
  and enables this bit to be reset to zero.
  Figure 11 presents the two part of the PREN operator
  used in the move multiple registers instruction.

244

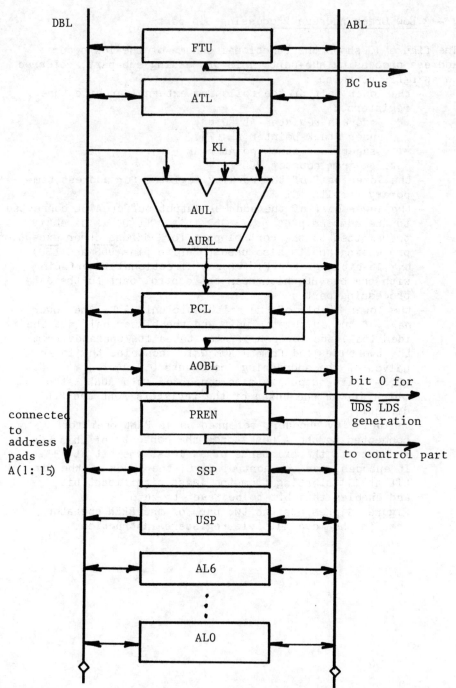

Figure 10 : Functional scheme of the lower order address processing
sub-part.

245

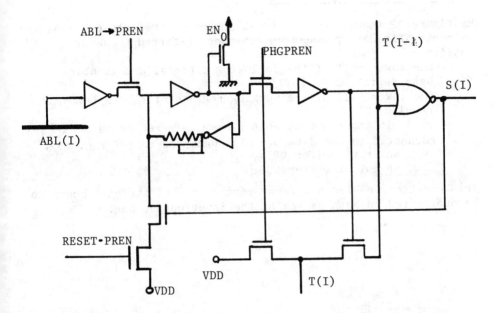

Figure 11 : PREN (register part).

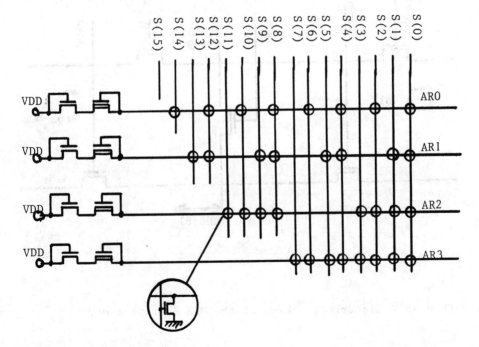

Figure 11 : PREN (encoder part).

246

1 - 4 Data Processing Sub-part

The figure 12 shows the functional scheme of the data processing sub-part. This data processing sub-part, referred to as DPP, contains :
- the lower half of the following external architecture data registers DL [0:7],
- the lower half of the working register for data temporary results DTL.
- a 16-bit input-output multiplexer referred to as MUXI/O connected to the data D(0 : 15) pads and to (cf fig. 13):
  . an input buffer DBIN,
  . an output buffer DOB.

This operator enables a byte present on one of the half buses to be duplicated on both halves of the input/output bus.

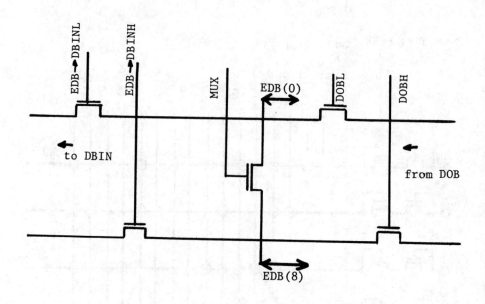

Figure 13 : Electrical scheme of the MUX I/O operator.

247

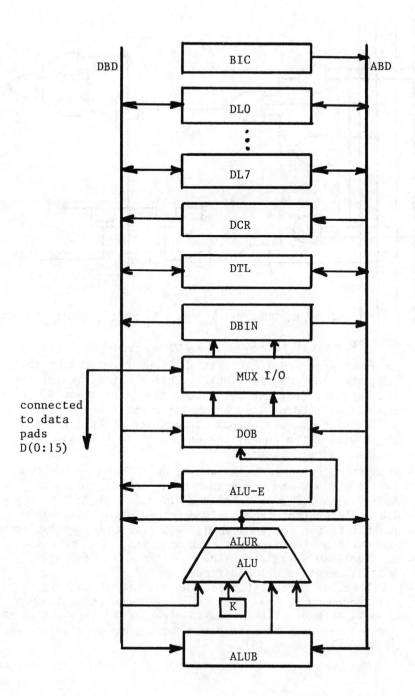

Figure 12 : Functional scheme of the data processing sub-part.

248

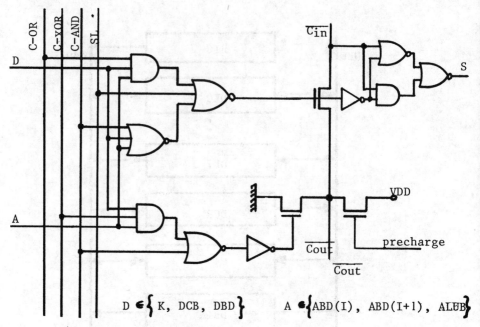

Figure 14 : Electrical scheme of the ALU operator.

- an arithmetic and logic unit which performs the operations :
  ADD, SUB, AND, OR, XOR, SHIFT. This ALU (cf fig 14) takes
  its operands from the two buses, from a constant ROM, from
  the buffer ALUB used as accumulator and from the decimal
  adjust circuit. Its output is connected to a latch ALUR
  which drives the buses, the output buffer DOB and the
  decimal adjust circuit.
- a shifter, referred to as ALU-E, which performs shifts
  over the bus D and hence enables in conjunction with
  ALU: 32-bit shifts or rotations (cf fig. 15).
- a byte insertion circuit BIC combined with the A bus
  amplifiers for reprecharching, while writing, the upper
  eight lines of the bus so that only the lower byte is
  written into a register (the upper byte is read again
  and is therefore not modified).
- a bit access circuit for setting one bit of bus D to
  '1' according to the value of the 4 lower bits(bit⟨0=3⟩)
  of busA (bit 4 of this bus is directly transmitted to the
  control part for the lower word or higher word selection
  as shown in figure 16).

249

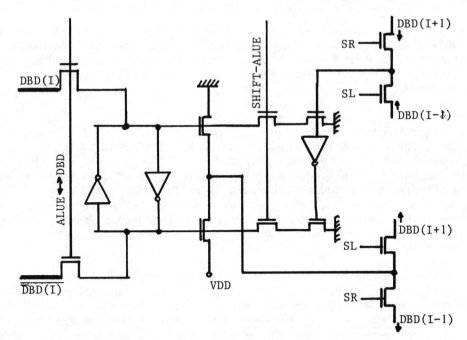

Figure 15 : Electrical scheme of the shifter ALU-E.

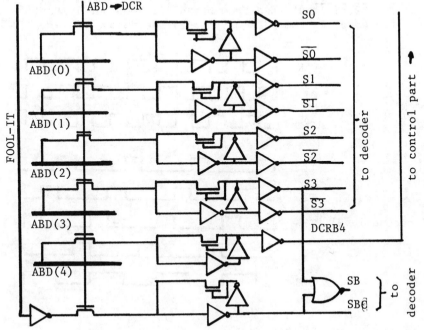

Figure 16 : Electrical scheme of the DCR operator (reg. only).

1 - 5 General Features of all three Processing Sub-parts

1 - 5.1 Carry look-ahead. The arithmetic and logic units (AUH - AUL - ALU) of the three sub-parts use a carry look-ahead mechanism on groups of 4 bits. A distributed NOR gate directly generates the carry from a group, when its 4 cells all propagate the carry.

1 - 5.2 Bus system. Two dual buses (bifilar bus) A and D run throught the whole data processing section. Two sets of switches enable these processing sub-part to work independently, in parallel or linked for data transfers.
For instance,the processing of the upper halves of data stored in the high order address sub-part is performed by the data processing sub-part. These dual buses behave like static memory buses, that is, a transfer is carried out in three phases : bus precharge, read (source connection), write (destination connection).

1 - 5.3 Clock system. The MC 68000 receives a continuous clock signal from an external oscillator (TTL level). Two sequencing levels are generated from this external clock signal :
    - first level : two non-overlapping phases ($\emptyset$1 and $\emptyset$2) are generated from the external clock at the same frequency,
    - second level : four successible instant signals T4, $\overline{T1}$, $\overline{T2}$, $\overline{T3}$ generated from $\emptyset$1 and $\emptyset$2. The four instants frequency is the half of ($\emptyset$1,$\emptyset$2) frequency (cf fig. 17)

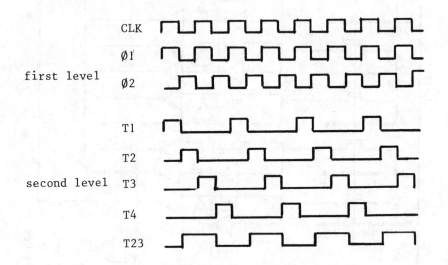

Figure 17 : MC 68000 Clock system

The sequencing of the data processing section is hence given
by the following rules (no system clock stopping due to external
exercising of HALT pins, or DTACK and so on...).
Each bus of each sub-part is able to perform a transfer between
two storage elements or an operation on a T4 through T3 cycle.

Bus transfer :
    - T4 : bus precharge
    - T1 : read source register (amplifiers triggered after 10ns)
    - T2-T3 : bus held ty the sense amplifier
    - T3 : write to destination register(s)
Operation :
    - T4 : precharge of dynamic structures (carries)
    - T1 : inputs become valid
    - T1T2 : calculation
    - T3 : storage of results

2 - CONTROL SECTION

2 - 1 Introduction

The control part of the MC 68000 microprocessor occupies approximately 70% of the chip area. Its functional organisation
is very classical, being of the parameterised microprogramming
type. The microprogram is stored in a 32K-bit ROM. A set of
PLAs performs decoding, conditional branching and exception
managing. The micro-instruction words are long (85 bits) and
consist of a 17-bits sequencing field and a 68-bits action
field which controls the data processing section through a
set of parametrisation PLAs actived by the operation word.
This set of PLAs and the logic circuits which generate the
commands form the most irregular part of the chip.
The figure 18 shows the functional scheme of the MC 68000 control
part. The control part contains the processor status word
(PSW) including system part (5 flip-flops) and user part (5 flip-
flops of condition code)

2 - 2 Microprogram ROM

The figure 19 shows the organisation of the MC 68000 microprogram
ROM.
The microprogram ROM is addressed by a 10 bit micro-address
which hence enables an addressing range of 1024 micro-words.
(for 544 actual micro-words). It is therefore possible to increase
the ROM contents without any modification of the external ROM
structure (address decoders, access, sequencing...)
The ROM is divided into two sections :

252

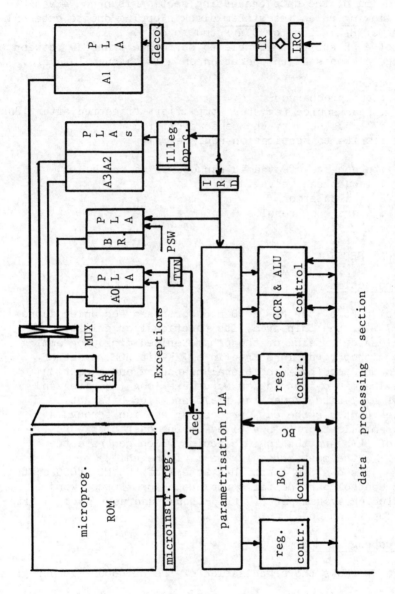

Figure 18 : MC 68000 functional block diagram of the control section.

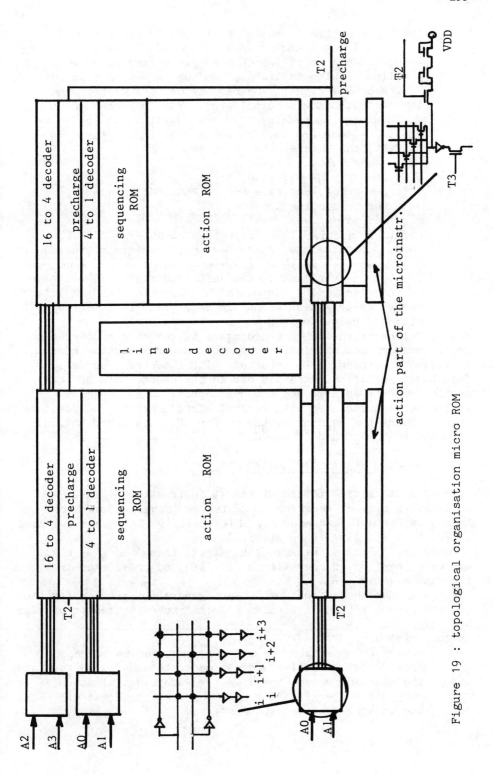

Figure 19 : topological organisation micro ROM

- the upper section : 34 (40 available) x 272-bit words, contains the sequencing part of the microinstructions (544 x 17-bit words) which are multiplexed to 16 words per line (8 bits multiplexed on the left-hand side and 9 bits on the right). The reading of this part of the ROM activates the sequencing mechanism situated above. Although the sequencing field has 17-bit width, only 10-bit of them are used to get a microinstruction address. In fact, the 7 more bits precise :
    . a branch occurence,
    . the branch condition,
    . the processor state (drive FC0, FC1 pads)
    . the sequencing type,
    . and so on...
- the lower section : 84 (over 88 available) x 272 bits words, contains the action part of the microinstruction (336 x 68-bit words) multiplexed to 4 words per line (32 bits multiplexed on the left-hand side and 36 bits on the right). The outputs of this part of the ROM activate the parametrisation mechanism data processing section situated below the ROM.

Both sections of the ROM are addressed by the same microaddress. Their decoders and matrices are placed back to back in order to implement a single ROM structure with a central decoder. The absence of certain transistors in the decoder of the lower section results in some address overlaps allowing the same action to be associated with several microinstruction addresses, this explains the reason why the number of action-word is lower than the number of address-word.

2 - 3 Decoding PLA A1, A2 and A3

These PLAs take the 16 bits of the IR instruction register (and the 16 bits of their complements) as inputs and each of them provides a 10-bit address. Electrically PLA A1 has switched loads while PLA A2-A3 have static loads.
PLAs A2 and A3 share the same AND matrix, these two PLAs also generate the invalid operation code (IOC) and privileged instruction signals (PRIV).
The PLA A1 decode the instruction and generate a microinstruction address except if the "effective address" mode is used to provide the operand address.
In such case :
    - either there exists one and only one operand address provided by effective address mode, though PLA A1 generates the microaddress of the effective address evaluation subroutine then PLA A2 provides the microaddress of the instruction execution program.

- either there exist two effective address modes provided
  operands, and then, PLA A1 generates the microaddress
  of the source operand effective address evaluation subroutine, then PLA A2 generates the microaddress of the
  destination operand effective address evaluation subroutine. Afterwards, PLA A3 provides the microaddress
  of the execution microprogram.

The figure 20 represents PLAs participation in the microprogram
execution.

2 - 4 Exception PLA A0

This PLA generates a microaddress when an exception condition
occurs. The possible exception conditions are :
- reset,
- BERR pin activation (bus error),
- ADERR internal generation (address error),
- occurence of M6800 mode interrupt,
- occurence of a vectored interrupt,
- occurence of a bus error before and existing interrupt
  exception is processed,
- interrupt recognition (output of interrupt logic)
- an attempt to execute an illegal instruction, detected
  by PLA A2, A3,
- trace mode while test signal being set,
- detection of op-codes 1010 and 1111.

Upon receiving one of these exception signals, the PLA A0 provides
the microaddress of the exception microprogram treatment.
The exception is processed corresponding to its priority. Table I
gives the order of priority in handling exception, in decreasing
order of priority.

256

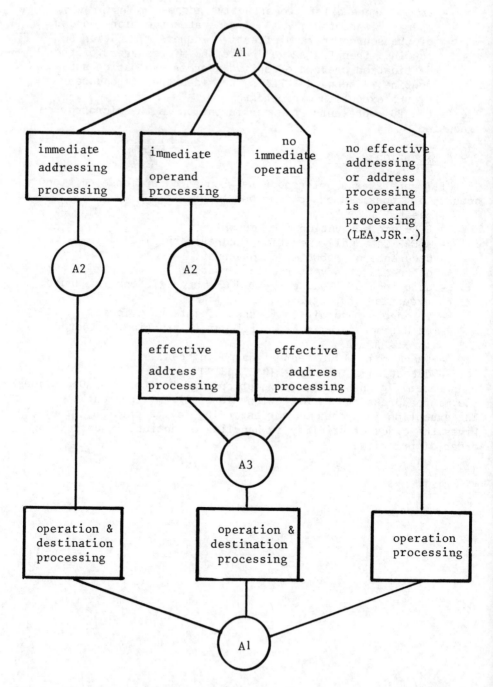

Figure 20 : PLA participation in the microprogram execution.

GROUP	EXCEPTION	PROCESSING
0	Reset Bus Error Address Error	Exception processing begins at the next minor cycle
1	Trace Interrupt Illegal Privilege	Exception processing begins before the next instruction
2	TRAP, TRAPV CHK Zero divide	Exception processing is started by normal instruction execution

Table I : decreasing order of priority of MC 68000 exceptions

2 - 5 Conditional branching PLA

This PLA, selected by one format of microinstruction, is used to generate the two least significant bits of the microaddress from the contents of the microinstruction sequencing field which provides the condition to be verified on one hand, and from the internal machine conditions for evaluation.
In this PLA, the AND plane is devided into two parts, the first part evaluates the condition while the second part, activated by a special shape in the first part, decodes the type of condition and performs the evaluations of the condition. The second part corresponds to Bcc, DBcc and Scc instructions where the condition is coded inside the operation code.

2 - 6 Parametrisation PLA

This is a very long PLA which appears as a folded strip above the data processing section.
Its is actived by a register IRD (cf fig.18) which is loaded

from the instruction register. From the operation word, this
PLA can calculate up to 4 possible commands (parameters) for
each operators of the data processing section. These parameters
are selected by the action part of the microinstructions and
are concerned with :
- register control (high order and low order in each sub-
  part of the data processing section) using the register
  number contained in the op-code,
- ALU control managing at the same time the operator commands,
  the operand acquisition and condition code (cf figure 20),
- ALU-E control in conjunction with ALU control,
- BC bus control and all operations using the BC bus for
  transfers between control part and data processing parts.

The register and ALU are not controlled directly by the parametrisation PLA but via command PLAs which generate the activation
conditions of the control lines depending on the type of commands
required.
The parametrisation PLA generates about forty different commands.
For instance, it enables the ADD, ADDX, SUB, SUBX, AND, OR
instructions to be handled by a single microcode sequence.

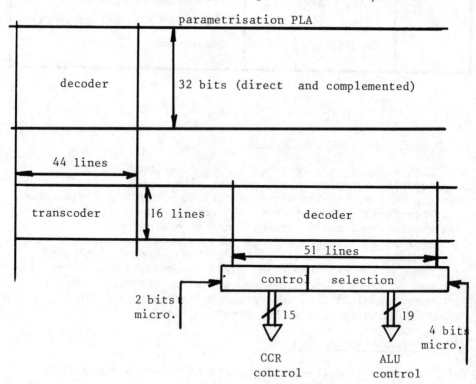

Figure 20 : command generation for the data calculation ALU

2 - 7 Sequencing of the Control Section

The figure 21 shows the sequencing of the control section.
The action part of a microinstruction is valid for a complete
instant cycle between two T3 signals. The T3 signals activate
the microinstruction register load. Two instant cycles are
then necessary before the validation of the desired microinstruction.
The sequencing of the microprogram ROM is resumed by :
- T1 : loading of bit <9:8> <5=0> of the microaddress
   register (column demultiplexer),
- T2 : column demultiplexing
      loading of bits <7:6> of the microaddress register
   (coming from branching PLA perhaps),
- T3 row demultiplexing
   microinstruction loading.

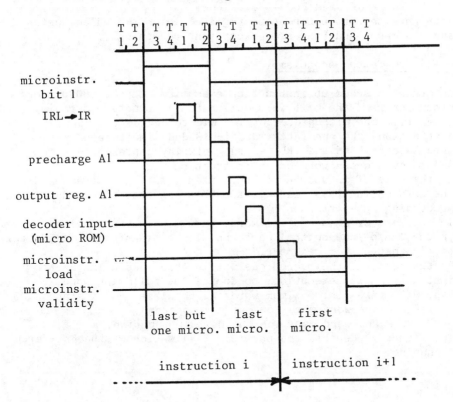

Figure 21 : sequencing of the control part

260

2 - 8 Status Register

The status register PSW is located centrally.
The condition code part is loaded directly from the ALU of
the data processing sub-part or from the BC bus which allows
exchanges with the external memory.
The system status register contains the 3 flip-flops of the
interrupt level, the S flip-flop for the supervisor/user mode
and the T flip-flop for trace mode. The interrupt level flip-
flops are managed by the interrupt logic when an interrupt
of higher priority occurs. The S and T flips-flops are managed
either by the processor itself or by the privileged instruction.
The static register of the physical processor contains the
following information:_:
- state of pin R/$\overline{W}$,
- exception processing NE,
- state of pins FC0 - FC1 - FC2.
This physical processor state register is only handled during
address error and bus error exception processing.

2 - 9 Pre-fetch Mechanism

In order to speed up instruction execution, the MC 68000 microprocessor read the next word during the instruction decoding.
Generally, the microprocessor is able to decide the correct
destination; the pre-fetch word is loaded simultaneously
into latches DBIN and IRL, as the word may represent either
a data or address word or an instruction.
In the case of an operand or the extension of an operand address,
the information is already is the DBIN register of the data
processing sub-part when it is required.
In the case of an operation word, the information is already
in the IRL of the control part and can be transferred immediately
to the instruction register.
Furthermore, if the microprocessor is able, depending on the
instruction step execution, to decide the following word(s)
type, it takes the maximum words as possible. For instance :
- the 1st word more for the next operand,
- the 2nd word more for the next instruction.
The figure 22 depicts the physical realisation of the pre-fetch
mechanism.

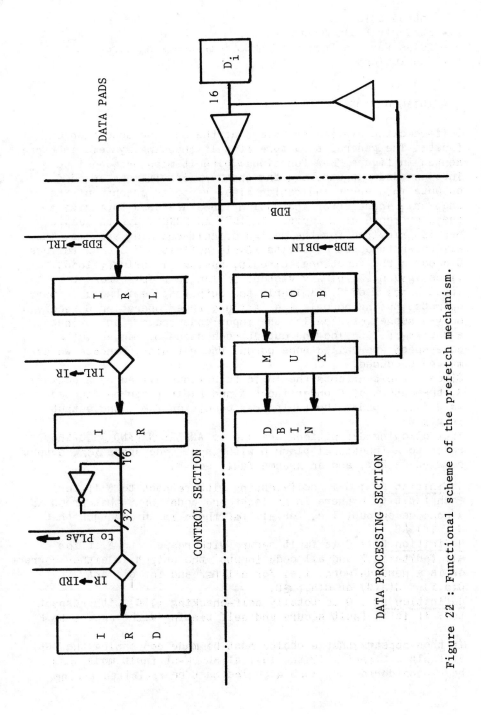

Figure 22 : Functional scheme of the prefetch mechanism.

PART C - DESIGN OF NMOS SELF-CHECKING CIRCUITS

1 - INTRODUCTION

Self-checking systems are aimed at the on-line detection of
faults. The general structure of self-checking systems is represented in figure 23. A functional circuit maps encoded inputs
into encoded outputs, and the checker circuit maps encoded
outputs into error indication lines (encoded outputs of the
checker). These error indication lines are used to indicate
the occurrence of a fault. In 1971 ANDERSON gave the basic properties required for the design of self-checking systems. They
all refer to a set of faults. Basic definitions are as follows:
G denotes the functional circuit, having r primary input
lines, giving 2^r binary vectors of length r, that form the
input space X of G. Similarly the output space Y has q primary
outputs. During normal, i.e. failure free, operation G receives
only a subset of X called the input code space A and produces
a subset of Y called the output code space B. Members of a
code space are called code words. Under faults, noncode words
may be produced.
Since a fault affects the logic function realized by a network
G, the output of G under input X and fault f can be denoted
as G (x,f). Under the fault free condition f=∅ , the output
is g(x,∅).
The following definitions are due to ANDERSON [AND 71]. They
refer to a functional block G with input code space A⊂X, output
code space B⊂Y, and an assume fault set F.

definition D1 : G is self-testing with respect to F if for
each fault in F there is at least one code input that produces
a noncode output, i.e. for all f∈F there is an a∈A such that
G (a,f)∉B.
definition D2 : G is fault secure with respect to F if for
all faults in F and all code inputs the output is either correct
or is a noncode word, i.e. for all f∈F and for all a∈A
G(a,f) = G(a,∅) or G(a,f)∉B.
definition D3 : G is totally self-checking (TSC) with respect
to F if it is fault secure and self testing with respect to F.

It then appears that a choice must be made concerning the set
of faults. Classically the logical stuck-at fault model has
been considered, but such a choice must be revisited taking

into account recent studies showing that the classical logical
stuck-at fault model might not be convenient to represent failures
that can occur in integrated circuits. This has been highlighted
by GALIAY et al. in [GAL 80] for single channel MOS technology.
Fault hypotheses for this technology may be defined [COU 81].
In the following, it is considered that we wish to design a
self-checking MC 68000, taking into account the following fault
hypotheses (class I) :
- all transistors stuck-on/stuck-open,
- failed contact or precontact,
- cuts of aluminum, diffusion or polysilicon lines,
- shorts between one aluminum line and the other nearest,
  (geographically) one ; idem for diffusion lines.

Such fault hypotheses reflect real faults that can occur, but
they imply to consider that the circuit may be redundant with
respect to some faults (e.g. a short between two aluminum
power lines). Hence a circuit could not be self-testing, and
hence could not be totally self-checking. Fortunately, the
basic properties given by ANDERSON have been extended in 1978
by SMITH-METZE. These authors defined the Strongly Fault Secure
circuits.

2 - STRONGLY FAULT SECURE CIRCUITS

Strongly fault secure (SFS) circuits have been introduced by
SMITH and METZE to overcome the difficulty introduced by circuits that are not totally self-checking (TSC) due to the loss
of self testing property. Given a network, known only to satisfy
the fault secure property for F there could be some fault $f1 \in F$
that could go undetected, and eventually a second fault $f2 \in F$
may occur. Then the fault $<f1, f2>$ is present, but $<f1, f2>$
may not be in F. Hence, there is no assurance that a code input
cannot cause the output to be an incorrect code word. As a
mater of fact, for a fault sequence $<f1, f2, ..., fn>$, a circuit may be fault secure with respect to any initial subsequences
and it is self-testing for the combination of faults in the
sequence. This has been formalized by the following definitions
[SMI 78] :

definition D4 : for a fault sequence $<f1, f2, ..., fn>$, let
k be the smallest integer for which there is an $a \in A$ such that

$$G\left(a, \bigcup_{j=1}^{k} fj\right) \neq G(a, \emptyset)$$

If there is no such k, set k=n. Then G is strongly fault secure
(SFS) with respect to the fault sequence if for all code words
$a \in A$

either $G(a, \bigcup_{j=1}^{k} fj) = G(a,\emptyset)$ or $G(a, \bigcup_{j=}^{k} fj) \notin B$

definition D5 : The network G is strongly fault secure (SFS) with respect to the fault set F if G is SFS with respect to all fault sequences whose members belong to F.

It is easy to see that every TSC network is SFS (k=1 in definition D4).

The problem now when dealing with SFS circuits as defined by SMITH-METZE and fault hypotheses as considered above at the level of transistors is that the number of faults is greatly increased than when higher level fault hypotheses are considered. If considering class I fault hypotheses, the list of possible faults for a circuit would be very long, if such a list could be done (consider that number of possible shorts between m lines increases as the square of m when the size of a circuit increases). Hence, an extended definition of SFS circuits may be given, using a generic class of fault hypotheses, and not a list of faults.

Prior to the definition of SFS circuits for a class C of fault hypotheses, redundant and strongly redundant circuits need to be defined.

definition D6 : A network which realizes the function $G(a,\emptyset)$ is redundant with respect to fault f and input code space A (resp. input space X), if

$G(a,f) = G(a,\emptyset)$ $\forall$ a$\in$A (resp. $G(a,f) = G(a,0)$).

definition D7 : A network is strongly redundant with respect to the fault sequence $<f1, f2, ..., fn>$ and with respect to the input space X (resp. input code space A) if the network is redundant with respect to the n fault sub-sequences $<f1>$, $<f1, f2>$, $<f1, f2, f3>$ ... $<f1, f2,..., fn>$ and with respect to the input space X (resp. input code space).

Now SFS circuits for a class C of fault hypotheses may be defined. Let C be a generic class of fault hypotheses. Assume a sequence of faults C will occur, the first being f1, the second f2, etc... Let $<f1, f2, ..., fk>$ the first fault sequence for which the network is not redundant. The network is redundant with respect to all fault sequences $<f1>$, $<f1, f2>$, $<f1, f2,..., fk-1>$.

definition D8 : G is SFS for the class C of fault hypotheses if for all sequences of faults f1 belonging to C which can occur, either :

     - $\exists$ k such that :

        . G is strongly redundant with respect to fault sequence $<f1, f2, ..., fk-1>$,

        . for all a$\in$A :

           either : $G(a,<f1, f2, ..., fk>) = G(a,\emptyset)$

or $G(a, <f1, f2, \ldots, fk>) \notin B$

. $\exists\ a \in A/G(a, <f1, f2, \ldots, fk>) \notin B$;

or :

- G is strongly redundant with respect to the fault sequences.

Although being the real definitions to be used for the design of self-checking integrated circuits, those definitions are not easy to handle. This is the reason for which general rules for the design of NMOS self-checking circuits based on the fault hypotheses given above, have been given. These rules consider different classical codes encoding the outputs of the functional circuit represented in figure 23. These codes are to detect single errors, or unidirectional errors, or multiple errors at the outputs of the functional circuit.

3 - RULES FOR THE DESIGN OF NMOS SELF-CHECKING CIRCUITS

In the following are derived errors due to faults belonging to the fault hypotheses (class I) considered above. The next examples are representative errors generated by these faults.

1 A signal diffusion line cut → a single error at the output of a gate or the concerned line.

2 A short between a signal diffusion line and the nearest line of the same type → one error which may concern the two lines but one and only one at a time → a single error.

3 A power aluminum line cut → multiple unidirectional errors at the output of gates powered by the line.

4 A short between 2 Vss or 2 Vdd power lines → no error.

So all cases of errors may be grouped into 4 categories, listed in the following :

- A : fault hypotheses giving single errors, concerning one signal line or one gate output, (1)
- B : fault hypotheses giving errors which may concern two signal lines/gate outputs, but one at a time, i.e. single errors, (2)
- C : fault hypotheses giving multiple but unidirectional errors, (3)
- D : fault hypotheses giving no error. (4)

In the following are considered rules to be used for the design of NMOS self-checking circuits to be checked for single errors at the outputs of the functional circuits (in [NIC 84-1] are also connected rules to be used for the design of NMOS self-checking circuits to be checked for unidirectional errors and multiple errors at the outputs of the functional circuits).

Rule 1: The maximum divergence degree of the functional block
with respect to the set of all internal lines belonging to
the functional block is equal to 1.

Rule 2 : One primary output of the functional block may be
erroneous due to multiple unidirectional errors created by
single fault hypothesis leading to C-type errors.

Rule R2' : Each diffusion and aluminum power line of the block
is used to power gates connected to only one primary output
of the block.

Rule R2" : Only one power (diffusion or aluminum) line is
used for VSS, and only one power (diffusion or aluminum) line
is used for VDD. The end of these lines is used to power the
gates of checker.

Proposition P1 : A circuit for which rule R2' is true is a
circuit for which rule R2 is true.
Proposition P1 holds simply since multiple unidirectional errors
due to a single fault hypothesis class I may come uniquely
from cuts of power diffusion or aluminum lines.

Proposition P2 : A circuit for which rule R2" is true is a
circuit for which faults belonging to class I and creating multiple unidirectional errors, can be removed from fault hypotheses.
Hence, rule R2 is true.

The group B and D of fault consequences derived in 3 may be
refined as follows :

B1 : B faults for which the two internal lines are such that
     if there were an error on both lines, only one primary
     output may be affected.

B2 : B faults for which the two internal lines are such that
     if there were an error on both lines, two primary outputs
     may be erroneous. Note that no more than two primary outputs
     may be affected due to Rule 1.

D1 : D faults for which the two power lines are powering gates
     whose outputs are such that all paths from outputs are
     towards one and only one primary output .

D2 : faults for which the two power lines are powering gates
     whose outputs are such that there exist paths from outputs
     to more than one primary output.

Let a sequence of faults <f1, f2, ..., fk-1, fk> for which
the circuit is
     - strongly redundant with respect to the fault sequence
       <f1, f2, ..., fk-1>,
     - not strongly redundant with respect to the fault sequence
       <f1, f2, ..., fk>.
According to the groups A, B1, B2, C, D1, D2 to which faults
f1, ..., fk belong, such a fault sequence may be denoted as
follows using regular expressions : (A+B1+B2+C+D1+D2)*.

Proposition P3 : If defects of type D2 may occur, then the circuit may not be SFS.

Proposition P3 may be illustrated simply by an example. The D2 type defect which occurred may leave the circuit in the situation depicted in figure 24.

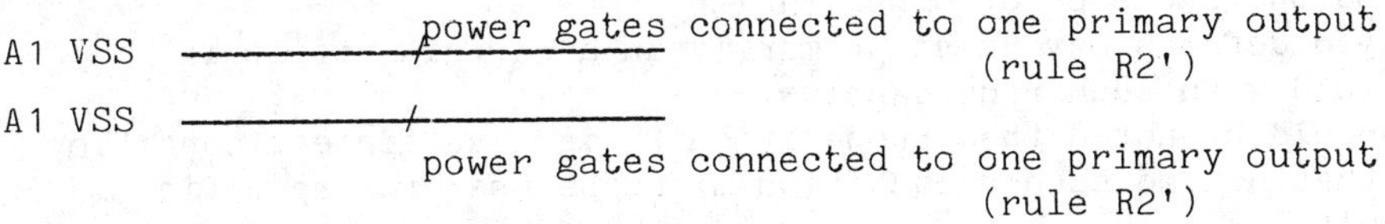

A1 VSS ——————————/———————— power gates connected to one primary output (rule R2')

A1 VSS ——————————/———————— power gates connected to one primary output (rule R2')

(a) initially (no fault)

A1 VSS ——————————┬————————
A1 VSS ——————————┴————————

(b) a D2 type defect for which the circuit is redundant with respect to the input space, occurs.

A1 VSS ——————————┬————————
A1 VSS ———/——————┴————————

(c) a cut (A-type defect) occurs. The circuit remains redundant.

A1 VSS ———/——————┬————————
A1 VSS ——————/———┴————————

(d) two primary outputs may be erroneous due to the successive occurrence of cuts on Aluminum lines.

Figure 24 : illustration for proposition P3.

Proposition P4 : If a subsequence <f1, f2, ..., fk-1> such that : the circuit is strongly redundant with respect to input code space and the type of faults is $(A+B1+C)^*$ B2, then the circuit may not be SFS.

Proposition P4 may be illustrated by a situation similar to the situation used for proposition P3.

Proposition P5 : If defects of types B2 and D2 cannot occur, then a circuit which has been designed of an output space code being checked for single errors, assuming rules R1 et R2', is SFS with respect to the generic class I of fault hypotheses. The fault sequence < f1, f2, ..., fk > will be such that : $(A+B1+C+D1)^*$.

Proposition P5 holds because the divergence degree of the circuit cannot be increased due to faults since the only types of defects which might increase the divergence degree are the B2 type and D2 type faults.

<u>Remarks</u> : Redundancy properties of a fault being dependent on the presence of other faults, it cannot be tolerated to have a present B2 type faults for which the circuit is not redundant when this fault is the only one present. The reason is that such a B2 type fault may give redundancy properties due to the presence of other faults.
D2 type defects may exist primarily in a circuit, if it is designed with such redundancies.
It should be noted that proposition P5 is a sufficient condition, and that not so strong condition might be used for specific circuits.

<u>Proposition P6</u> : If defects of types B2 cannot occur, then a circuit which has been designed for an output space code being checked for single errors, assuming rules R1 et R2", is SFS with respect to the generic class I of fault hypotheses. The fault sequence <f1, f2, ..., fk> will be such that : (A+B1+C)*.

<u>Rule 3</u> : To make a design such that B2 type defects cannot occur, it is necessary to list all shorts between internal lines (diffusion or aluminum) such that if there were an error on each line, two outputs could be erroneous. For such shorts, the corresponding lines have to be remoted or they have to be implemented with materials such that these shorts cannot occur. For example, one line has to be implemented with diffusion, and the other one with aluminum or polysilicon.

<u>Rule 4</u> : To make a design such that D2 type defects cannot occur, it is necessary to list all shorts between two power lines of the same type (VSS or VDD) and to eliminate all these possible shorts by making one line in diffusion and one line in aluminum, or by remoting these lines one from the other.

4 - CONCLUSION

Similar rules have been defined for the design of NMOS functional part of self-checking circuits whose outputs are encoded such that unidirectional or multiple errors are detected. The choice of these codes depends on the functionality of the circuit.
The next part exemplifies such choices for the design of a self-checking version of the MC 68000. A last remark is to notice that the design of checkers must be considered. Properties and design of checkers is the scope of [NIC 84-2]. These properties are not detailed here, but the surface overhead of those checkers will be taken into account in the next part.

PART D - EVALUATION OF A SELF-CHECKING VERSION OF THE
MC 68000

1 - Self-checking version of the data processing section
2 - Self-checking version of the control section

General layout rules similar to those given in part C are used
for the design of a self-checking version of the MC 68000 as
described in part B. The final result is that a strongly fault
secure version with respect to the fault hypotheses given in
part C yields in a total increase of only 48%.

1 - SELF-CHECKING VERSION OF THE DATA PROCESSING SECTION

The self-checking version for the data processing section of
MC 68000 is presented in figures 26, 27, 28.
The BUS, the RAM, the Latches, the Buffers, the Program Counters
(low and hight) the Stack Pointer, are all tested by a parity
code, since the layout rules given to design self-checking
circuits for single error detecting codes are checked by the
bit slice structure of these blocks.
Two parity bits are added to code the low byte and the hight
byte of these circuits since data may be transferred using 8
bits data width or a complete 16 bits data width.
All these circuits are controlled during transfer phase of data
by the parity checkers connected on the BUS of the data processing section.
Most of the operators (arithmetic and logic unit (ALU) extension
of arithmetic and logic unit (ALUE), low arithmetic unit (AUL),
hight arithmetic unit (AUH) and the priority encoder register)
are duplicated, since the analysis of faults gives that primary
output errors are multiple errors. Every duplicated block is
tested by a double-rail checker. Two parity bits are generated
before the results of these blocks are transferred on the bus.
The operators i.e. byte insertion circuit (BIC), decoder register
(DCR), sign extension circuit (SIGN-EX) and input/output multiplexer are tested by two parity bits, since the analysis of
faults gives that errors to primary outputs are single errors.
All these circuits are checked during transfer phase of data
by the parity checkers connected on the buses of the data processing section.
Lastly address output buffers and pads are checked by one parity
bit and data input/output buffers and pads are checked by two
parity bits since data may be transferred using 8 bits data
width or the complete 16 bits data width.

Figure 25 : presents the surface occupied by the data processing
section, the instruction register (IR and IRL), the channel
routing and by data and address   buffers and pads. Table II
gives a summary of area allocation for duplicated blocks of
data processing section, for blocks testing by two parity bits
for data and address buffers and pads and for the channel routing.
In the same table is given the area required for the self-checking
version of above blocks and for required checkers and parity
generators. It results that the area penalty required for this
solution is 45,5%.

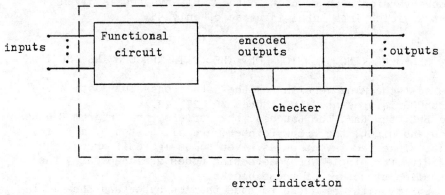

Figure 23 : General structure of self checking circuits.

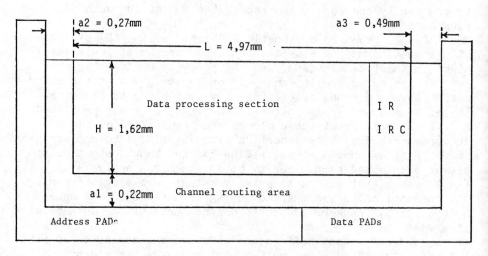

Figure 25 : Surface occupied by the data processing section.

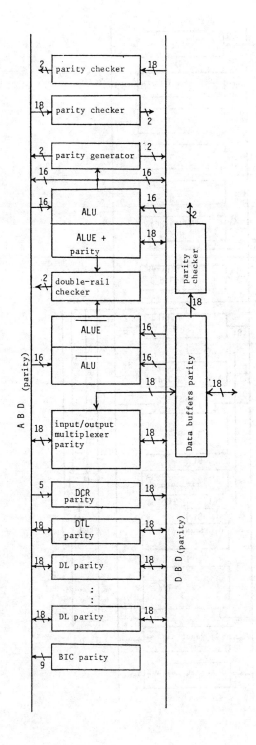

Figure 26 : block diagram of the self-checking data processing sub-part

272

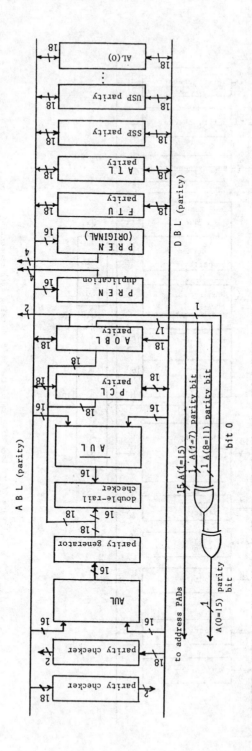

Figure 27 : block diagram of the self-checking low order address processing sub-part

273

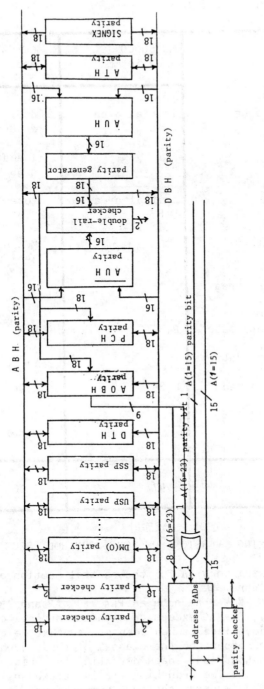

Figure 28 : block diagram of the self-checking hight order address processing sub-part

TABLE II

	MC 68000 area in mm^2	MC 68000 self-checking area in mm^2
duplicated blocks	initial area 2,18	duplicated block area 2,18 x 2 = 4,36 losed area 4,36 x 2/16 = 0,54
parity checked blocks	initial area 5,87	coded blocks 5,87 x 1 2/16 = 6,6
parity generators and checkers		MOS numbers x 1443 1467 x 1443 m^2 = 2,1
channel routing area (PO)	2,45	3,19
buffers and data pads	2,13	2,39
parity checker (databus)		0,14
Buffers address pads	2,84	2,96
parity checkers (address bus)		0,22
	15,47	22,50

2 - SELF-CHECKING VERSION OF THE CONTROL SECTION

The self-checking version for the control section of MC 68000 is given in figure 29.
In this version the PLAs such as PLA, A1, A2 and A3 (operation code decoding), PLA A0 exception decoding are all checked using the Berger code, since it has been proved that we can obtain -SFS PLA with respect to the class I of fault hypotheses, if we use an output code space detecting the unidirectional errors [NIC 84-1]. The branching PLA is checked using the double-rail code, since we need the same number (two) of suplementary

outputs to encode the two outputs of this PLA with the double
rail code or with the Berger code (see figure 29).
The analysis of faults for the microprogram ROM gives two error
types provoked by the defects of class I.

a/ The defects on the lines between the next address register
and the inverters of the lines decoder or the inverters of
the column decoder can generate the addressing of a micro-instruction instead of another one. It is obvious that such defects
cannot be detected by encoding of microinstructions. In the
self-checking version of micro-ROM (figure 30) these defects
are detected by a parity code. Some parity bits (P0, P1, P2,P3)
are generated at the outputs of next-address   register
and for all other sources of micro ROM address. So the address
lines A0-A9 are checked before the inputs of the inverters
of lines'decoder and columns' decoder.

b/ All other defects of the class I yields unidirectional errors
on the primary outputs of the ROM, due to the inversion parity
of paths between the internal lines of the ROM and the primary
outputs of the ROM. To detect such defects micro-instruction
are coded with Berger code (bits B0, B1, ... B6) and outputs
of the ROM are checked by a Berger code checker.

Finally a parity generator is used to generate some parity
bits necessary to control the transfer of the different fields
of the action part of the micro-instruction until the data
processing section.
The area penalty for this solution is 20% with a 100% coverage
of class I.
The parametrisation PLA is checked using a Berger code.
The different blocks which combine the bits of action field
and the outputs of the parametrisation PLA to generate the
control of the data processing section such as ALU and RCC
control, register control and so on... are implemented with
a random structure. The errors provoked on the outputs of
these blocks are multiple errors, therefore these blocks are
duplicated.
However it will be possible to use another cheaper method but
it will imply to redesign these blocks in order to ensure the
layout rules for the design of SFS circuits for as previously
said.

For the same reasons some blocks which are not presented in
figure 18 i.e. interruption logic, clock generation and Bus
control logic are duplicated.
Finally the invalid op. code detector is checked using the method
given in [NIC 84] for specific checkers, that is to say using

an automatic internal test pattern generation during unusual
cycle to insure the correct detection when it is used.
Table III gives pins count for both the simple and self-checking
microprocessors. Table IV gives a summary of area allocation
for the differents blocks of both simple and self-checking
microprocessor. The supplementary area needed for the self-
checking version is 21,56 mm² which represents a 48% area penalty.

TABLE   III

Initial pads	coding extension pads	self-checking circuit pads
data (16 pads)	parity (2 pads)	(18 pads)
address (23 pads)	parity (1 pad)	(24 pads)
exceptions :		
IPLO, IPL1, IPL2		
(3 pads)	parity (1 pad)	(4 pads)
function code :		
IFCO, IFC1		
(2 pads)	parity (1 pad)	(3 pads)
IFC2 (1 pad)	double rail (1pad)	(2 pads)
other control pads		
(14 pads)	double rail (14 pads)	(28 pads)
clock (1 pad)	duplication (1 pad)	(2 pads)
power supply :		
Vdd (2 pads)	duplication (0 pad)	(2 pads)
Vss (2 pads)	duplication (0 pad)	(2 pads)
0 error detection		
pad	double rail (2 pads)	(2 pads)
64 pads	23 pads	87 pads

277

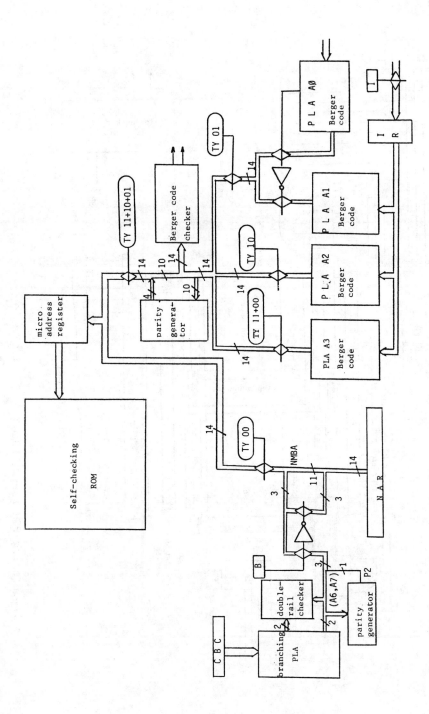

Figure 29: PLAs

278

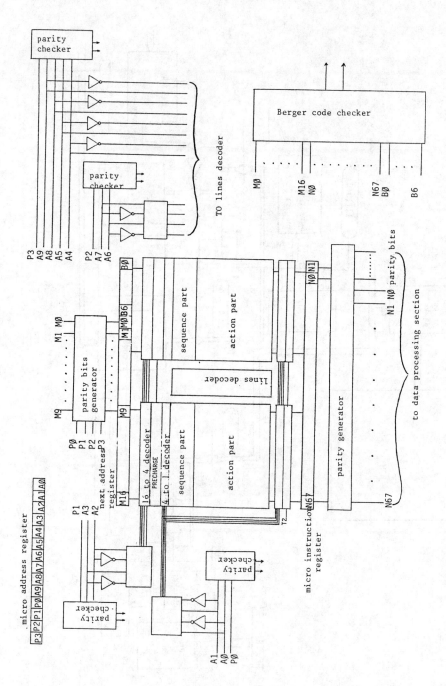

Figure 30: self-checking ROM for MC 68000

TABLE   IV

block	initial area	augmentation
data processing section	8,06 mm^2	5,54 mm^2
data pads	2,13 "	0,40 "
address pads	2,84 "	0,34 "
channel routing area (DPS)	2,45 "	0,74 "
ROM	8,17 "	1,64 "
PLAs A1, A2, A3 (Op.code decoding)	2,64 "	1,05 "
branching PLA	0,31 "	0,12 "
exception PLA	0,23 "	0,09 "
invalid op-code detection	0,55 "	0,81 "
parametrisation PLA	1,90 "	0,76 "
DPS control circuit	1,73 "	1,85 "
DPS control buffers	0,98 "	0,48 "
external bus control logic	1,1 "	1,17 "
exception circuitry	0,29 "	0,31 "
clock circuitry	1,89 "	1,81 "
pads other than address and data ones	3 "	2,76 "
channel routing area (PC)	6,34 "	1,60 "
error circuitry		0,09 "
	44,57 mm^2	21,56 mm^2

280

PART E - CONCLUSIONS

1 - Comparison with available results
2 - Feasibility of a self-checking version of the MC 68000
3 - Some specific behaviour of the MC 68000

Conclusions include three parts developed in the following.
The first one compares the final results obtained in part D to
the results given by DISPARTE in [DIS 81]. The second one deals
with the feasibility of a self-checking version of the MC 68000,
i.e. with the feasibility of a larger chip. The last one takes
the opposite view of sections C and D : although a great care
may be taken to face with numerous possible failures, it may
happen that without any failure, a misfunction occurs. This
is the case of the MC 68000, for which the behaviour does not
correspond to the specifications (but this is probably the
case for most available microprocessors).

1 - COMPARISON WITH AVAILABLE RESULTS

It is interesting to compare this study with the self-checking
version for MC 68000 microprocessor given in [DIS 81].
A first point of this comparison concerns the area penalty.
The present approach necessitates a 48% area penalty instead
of 58% area penalty for the approach of [DIS 81] .
The more important difference in these studies is that the
present approach is given to cover a well defined class of fault
hypotheses and the detailed structure of every block may be
known to find the convenient solution , when the study given
in [DIS 81] is based on existing studies for self-checking
microprocessors, which are modified and completed but some
blocks are not known in all details and some of them are duplicated when more economical solutions may give a suficient protection. Moreover, some blocks are tested by methods which
are not suficient to cover realistic   fault hypotheses.
For example the data processing section is duplicated when
a more economical solution may be used with sufficient coverage
of defects. External data BUS is checked by a parity bit when
at least two parity bits are necessary since data may be transferred using 8 bits data width or the complete 16 bits data
width. "Execution control" is checked using a parity code which
is a very economical solution but this block is composed by
a very long PLA (parametrisation PLA) and many circuits implemented with a random structure and the parity code may cover a
small portion of real defects. "Instruction decode" (PLA, A1,

A2, A3) is checked by regeneration from the "present micro-instruction address" and comparison between the regenerated op-code and the requested op-code.
But the recoding of the op-code from the "present microinstruction address" is not realistic since such a method necessitates to use an inverse circuit which exists only if the input code space and the output code space are bijective. In a microprocessor many different op-codes may address the same microinstruction address.
The parity code used in [DIS 81] to check the control field of microinstruction is not sufficient to cover all real defects.
Finally the ROM addressing is protected in [DIS 81] by storing in each ROM location a complete binary present address field.
The ROM addressing is completely protected by this method if in each ROM location a correct "next address field" is stored and if it is transferred correctly in the "next address register". But it is not protected againts defects modifying the "next address field" or the contents of the "next address register".
Another reason for which this method fails is that two separate column decoders and two separate line decoders are used to address the address field and the control field of microinstruc-tion address field.
It results that the micro ROM is not protected sufficiently in spite of a 40% area penalty needed by this solution. In the ofter hand a 20% area penalty may protect the micro ROM if a systematic method is used to analyse the errors on the micro ROM outputs due to real defects.
The interest to use such a method is shown also in [NIC 84-1] when it appears that general methods based on the functional description of microprogrammed units, presented in [TOY 78] and [NAM 82], is not sufficient to protect the micro ROM of MC 68000 againts real defects.

2 - FEASIBILITY OF A SELF-CHECKING VERSION OF THE MC 68000

It might be thought paradoxical to be able to build a CPU as complex as the MC 68000 and to be unable to circumvent the problem of building a self-checking version of that microprocessor. In fact the problem mooted by the self-checking version is twice : first the silicon area, secondly the pads number.
The technology (HMOS 1) used to build the MC 68000 microprocessor in 1978 has 3,5 micron minimum gate size for a total area of 44, 53mm^2 (6,1 x 7,3). Todays the available technology allows 2 mi-crons minimum gate size.
Taking into account the increase of scale integration for a dynamic RAM, as shown in table $_{\text{v}}$.

Year of product	minimum gate size (micron)	die size (mm)	chip/wafer (Ø 125mm)
1978	4	6,3x6,3	245
1981	2,5	5 x5	450
1984	1,7	4,3x4,3	580
1987	1,2	4 x4	670

Table V . Increase of scale integration for dynamic RAM.

The silicon area needed to design an MC 68000 microprocessor today, would be about 24,0mm^2 (4,6x5,2). This represents a benefit of about 20,53mm^2, that is to say 60% of the first MC 68000.
On the other hand, a self-checking version of the MC 68000 needs a 48% increase in silicon area. Taking today's silicon area (24mm^2), a 48% increase of this area will need 11,5mm^2 more for the self-checking part. So the whole area of a self-checking version of today would be 35,5mm^2 that is to say, smaller than the area gained by the increase of scale integration.
Note that the total area can be built without any problem. The second problem is not in fact a real problem. It is up to the area occupied by the unbedded chip. A current 40 pins package is 51mm long, the MC 68000 is, for the moment, the longest chip in term of its package: 64 pins package, that is to say 82mm long. The self-checking version with its 88 pins would need a 113mm package. This is not impossible to build, but it is very large to populate a board. So, perhaps, the implementation of the self-checking version with a square package such as is used for embedding hybrid circuit would be better. The size would then be no larger than a current 40 pin-package because it will only need 22 pins of each of its 4 sides. Of course, a bed-nails package is also possible, but would be under-used.

3 - SOME SPECIFIC BEHAVIOUR OF THE MC 68000

The study of the internal microarchitecture of the MC 68000 leads us to observe the behaviour of that microprocessor in order to verify the fidelity between the data sheet description made by MOTOROLA and the actual processor.
Now the internal observation has allowed us to see that there exists some discordances between the MOTOROLA data sheet and what effectively happens in the microprocessor. These are explained below.

3 - 1 Definitions

Let us call RESET (hard), the assertion of the RESET signal by an external activation of the RESET pin from an external device.
Let us call RESET (soft), the assertion of the RESET by an internal activation of the RESET, this corresponds to the execution of the RESET instruction.
Let us call HALT (hard) : the assertion of the HALT signal by an external activation of the HALT pin from an external device.
Let us call HALT (soft) : the assertion of the HALT signal by an internal activation of the HALT pin. This corresponds to the execution of the HALT 'microinstruction' (T6E mask) or to the execution of the two HALT 'microinstructions' (DL6 mask).

3 - 2 Remind

The RESET (soft) is exercised in order to reset all the external device of the MC 68000, without any action over the microprocessor itself. This one stays in its state (instruction, execution).
The HALT (soft) is exercised in order to stop the microprocessor when a bus error or an address error occurs during exception processing that is to say bus error processing, address error processing, reset.
The RESET (hard) is exercised in order to reset the processor itself by the execution of the initialisation sequence.
The HALT (hard) is exercised in order to stop the microprocessor at the end of the current instruction exection bus cycle until the halt pin is desactivated. The processor keeps on then in normal mode. According to MOTOROLA data sheet:
The RESET (hard) signal allows the 'microprocessor reset' while simultaneous activation of RESET (hard) and HALT (hard) allows a 'total system reset' (CPU and external devices).

3 - 3 The Double Conflict of RESET (hard) and RESET (soft)

The two RESET signals : RESET (hard) and RESET (soft) are at strife. The first conflict is produced by both signals themselves : one masking the other. The second conflict is produced during HALT pin assertion.

3 - 3.1 RESET's conflict. Let us suppose a microprocessor in instruction mode, let the current instruction be a RESET (soft) instruction executed to reinitiate the external devices. This instruction is performed during 124*[Ø1, Ø2] cycles, that is to say 15,5 microseconds (for MC 68000 running at 8MHZ). During that time, the assertion of the RESET button in order to generate the RESET (hard) signal to reinitiate the microprocessor will not have any influence. Therefore, in spite of beginning the reset sequence, the processor will keep on the RESET (soft) instruction execution, and it will then keep on with the next instruction execution.

284

In this case, the only means to restart the system is to exercise HALT pin and RESET simultaneously.
Hence :
During the execution of the RESET (soft) instruction if the RESET pin is exercised : instead of starting the reset sequence, (exception sequence of highest priority) the microprocessor keeps on executing the RESET (soft) instruction and then the following ones and so on and so forth.
The behaviour is still the same if a set of RESET (soft) instructions is executed instead of only one. Resetting in microprocessor will not be possible by activating the RESET.

3 - 3.2 RESET and HALT conflicts. Let us suppose the same conditions as previously : that is to say the microprocessor in instruction execution mode, the current instruction is a RESET (soft) instruction. If during this instruction execution, the microprocessor receives an external activation of the HALT pin, an abnormal behaviour occurs. Effectively, instead of stopping (HALT) and waiting until the end of the external locking of the HALT (hard), the microprocessor begins the execution of the reset sequence loosing hence the current context.
Two origins are possible for the HALT (hard) generation :
- an external device connected on the external RESET line, in that case, this device generate the conflict if it activates the HALT before receiving the RESET signal from the microprocessor ;
- an external device not connected on the external RESET line but activating HALT signal in order to use the external bus of the MC 68000 (typically the case of a DMA device).
Hence :
During the execution of the RESET (soft) instruction, if the HALT pin is exercised :
instead of halting the CPU until the HALT signal goes down, and then executing the next instruction.
The microprocessor will stop instruction execution and it will start the system reset sequence, loosing the current context.

Remark :
We can remark that the modification of the starting address of the RESET routine : address 0 (T6E) to address 2 (DL6) in order to circumvent the problem of starting with an error, do not have any influence on these anomalies of the MC 68000 microprocessor.

REFERENCES

[AND 71] ANDERSON D.A.,
          "Design of self-checking digital networks
          using coding techniques."
          Coordinated Sciences Laboratory, Report R/527,
          University of Illinois, Urbana, Sept. 1971.

[BAS 84] BASCHIERA D., DERANTONIAN M., MARCHAL P., NICOLAIDES M.
          "Fonctionnement et test du MC 68000"
          Rapport final de contrat EDF n° 120IB 1455 Janvier 1984

[COU 81] COURTOIS B.
          "Failure mechanisms, fault hypotheses and analytical tes-
          ting of LSI-NMOS (HMOS) circuits"
          VLSI 81, University of Edinburgh, August 18/21 1981
          United Kindom Academic press.

[DIS 81] DISPARTE CH.
          "A design approach for an electronic engine controller
          self-checking microprocessor"
          EUROMICRO Symposium, Paris september 1981.

[FRA 80] FRANK E.H., SPROULL R.F.
          "An approach to debugging custom integrated circuits"
          CMU Annual report 1979-1980.

[GAL 80] GALIAY J., CROUZET Y., VERGNIAULT M.
          "Physical versus logical fault models MOS LSI circuits :
          impact on their testability"
          IEEE Trans. on comp. vol.C-29 n°6 june 1980.

[MAR 83] MARCHAL P.
          " Test en ligne du microprocesseur MC 68000. Modelisation
          et programme de test"
          These de 3ème cycle INPG Grenoble juillet 1983.

[NAM 82] NAMJOO M.
          "Design of concurrently testable microprogrammed control
          units"
          Technological report n°82-6 juin 1982
          Center of Reliable Computing, Standford University.

[NIC 84] NICOLAIDIS M.
          "Conception des circuits intégrés autotestables, pour des
          hypothèses de pannes analytiques"
          thèse de docteur ingénieur INPG janvier 1984.

[NIC 84-2] NICOLAIDIS M., JANSCH I., COURTOIS B.
" Strongly code disjoint checkers"
14th FTCS Orlando Florida 20/22 june 1984.

[SMI 78] SMITH J.E., METZE G.
"Strongly fault secure logic networks"
IEEE Trans. on Computers, vol C-27 n°6 june 1978.

[TOY 78] TOY W.N.
"Fault tolerant design of local ESS processors"
Proc. IEEE vol 66, pp. 1126/1145 October 1978.

SUBJECT INDEX